WEATHER FORECASTING
FOR SAILORS

Frank Singleton joined the Meteorological Office in 1955 and
has worked as a forecaster in the UK and the Middle East.
He is an enthusiastic dinghy sailor and has run many courses
on weather forecasting and sailing. He is co-author of *The
Yachtsman's Weather Map* which was published by the Royal
Yachting Association in 1975.

D0620295

TEACH YOURSELF BOOKS

To Jennifer and Jayne

WEATHER FORECASTING FOR SAILORS

Frank Singleton

Illustrations by J. R. Nicholas

TEACH YOURSELF BOOKS
Hodder and Stoughton

First printed 1981

Copyright © 1981
Frank Singleton

British Library C.I.P.

Singleton, Frank
  Weather forecasting for sailors. – (Teach yourself books)
  1. Weather forecasting
  2. Yachts and yachting
  I. Title
  551.6′3   QC995

  ISBN 0–340–25977–9

Filmset by Northumberland Press Ltd, Gateshead, Tyne and Wear. Printed and bound in Great Britain for Hodder and Stoughton paperbacks, a division of Hodder and Stoughton Ltd, Mill Road, Dunton Green, Sevenoaks, Kent, (Editorial Office; 47 Bedford Square, London, WCIB 3DP) by Richard Clay (The Chaucer Press) Ltd, Bungay, Suffolk

Contents

Acknowledgements

I wish to thank the Royal Meteorological Society and the Royal Yachting Association for permission to reproduce the METMAP. I am grateful to members of the Meteorological Office Camera Club for providing the cloud photographs and to Mr J. R. Nicholas for the artwork. I am also very pleased to record the help and advice given to me by the editorial staff of Teach Yourself Books.

Glossary

Air Mass A body of air in which the horizontal changes in temperature and humidity are relatively slight. The two main types are 'Polar' and 'Tropical' indicating the origin of the air.

Anabatic Adjective used to describe a local wind which blows up a slope heated by sunshine.

Anemometer Instrument for measuring wind speed.

Aneroid Barometer A 'without liquid' barometer – basically a partially evacuated metallic capsule surrounding a spring.

Bar Unit of pressure approximately equal to one of atmosphere, normally divided into millibars, the usual units.

Barograph A recording barometer – an aneroid capsule, the expansion and contraction of which actuates the movement of a pen to leave a trace on a chart wound around a revolving drum.

Barometer Instrument for measuring pressure, the original invention of Torricelli consisted of a long tube full of mercury.

Beaufort Scale	A scale of wind force using numbers ranging from 0, calm, to 12, hurricane force.
Cirro-stratus	Transparent whitish cloud of uniform or smooth appearance, partly or wholly covering the sky. Often formed, apparently, by the merging of cirrus streaks.
Cirrus	Detached clouds at high level in the form of delicate filaments, fibrous or hairlike.
Condensation	The process of formation of a liquid from its vapour. In the atmosphere the cooling of air leads to condensation of water from water vapour contained in the air.
Convection	The process of vertical motion created by buoyancy due to warming of air in the atmosphere; convection can, of course, occur in any fluid.
Coriolis effect	Because of the rotation of the earth, air moving from one place to another appears to an observer standing on the earth to be deflected. The deflection is to the right in the northern hemisphere (rotating counter-clockwise) and is to the left in the southern hemisphere.
Cumulo-nimbus	Large very well developed cumulus type cloud often associated with thunder and usually giving showers. The tops often spread out to form an anvil; this is at cirrus levels and can remain after the dissipation of the lower part of the cloud.
Cumulus	Detached 'heaps' of cloud formed by convection, usually forming at relatively low levels.
Cyclone	Atmospheric pressure distribution surrounding a central low pressure. In middle or high latitudes usually referred to as a 'low' or 'depression'.
Dew-point	The temperature to which air must be cooled for condensation to occur.
Drizzle	Liquid precipitation in the form of small drops of near uniform size falling from stratus cloud.
Foehn effect	The mechanism responsible for the warm dry wind which occurs to the leeward side of mountains or hills. (Also called Föhn.)

Fog	Small water droplets in suspension in the air at ground level, effectively a very low level cloud.
Hill fog	Low cloud enveloping hills. The cloud may be formed by the forced uplift as the air flows over the hill or may just be very low level cloud which happens to be over high ground.
Radiation fog	Fog which forms overland on a clear night with light winds.
Sea fog	Fog which forms over the sea when air blows from a warm area of water towards a colder one. Sea fog is a particular case of what the textbooks refer to as *Advection fog* – this can occur when air blows from a relatively warm area towards a cold one, say with lying snow or a frozen ground.
Front	The boundary between two air masses of different type. A front necessarily lies in a trough of low pressure and is marked by discontinuities in wind direction and velocity, humidity, temperature, weather and visibility. Precisely how marked will be the discontinuity varies greatly from one front to another.
Cold front	A front whose movement is such that colder air is replacing a warmer air mass.
Warm front	One where warmer air is replacing cold.
Occluded front (Occlusion)	An occluded front is one where the cold front has caught up with the warm for the two fronts effectively to merge.
Geostrophic scale	A scale relating the spacing of isobars to the wind speed.
High	Atmospheric pressure system surrounding a central high pressure. May be referred to as an anticyclone.
Blocking high	A high which is stationary and sufficiently large as to prevent the normal west to east movement of lows.

Building high	A high in which the central pressure is rising and the system probably increasing in size. May be described as intensifying.
Declining high	One where the central pressure is decreasing. May be known as weakening or, if the decrease in pressure is rapid, as collapsing.
Sub-tropical high	One of the highs which compose the quasi-permanent belt of high pressure of the subtropics (this is the part of the surface of the earth between the tropics and latitude 40 degrees).
Hurricane	Intense tropical cyclones which occur in the West Indies and the Gulf of Mexico.
Isobar	A line of equal pressure.
Jet stream	A strong narrow current of air at high levels in the atmosphere. In temperate latitudes the jet stream is almost invariably from a westerly point while to the south of the subtropical highs there is commonly an easterly jet.
Katabatic	The adjective used to describe the wind which blows down a slope which is cooling at night time.
Lenticularis	Adjective used to describe cloud having an ovoid or lens shape and usually associated with standing waves.
Local winds	The behaviour of winds which can be attributed to local, often topographical, effects rather than the large scale pressure pattern. Sea breezes, anabatics, katabatics, bends around headlands and gusts are examples of local effects which might either create winds in their own right or modify to varying extents the (large scale) pressure gradient wind.
Low	*See cyclone.* Low or depression is the more usual name.
Complex low	A low with two or more identifiable centres.
Deepening low	One where the central pressure is decreasing. May be referred to as intensifying.
Filling low	One where the central pressure is increasing.

Polar low

A small and often weak (in terms of wind strength) low forming over the sea to the north of Britain and not associated with any fronts. Particularly a winter phenomenon.

Monsoon

A seasonal wind which blows for about six months from the north-east and then six months from the south-west over the Arabian sea. Similar effects occur in some other parts of the world but are much less marked.

Psychrometer

Instrument for measuring the dew-point and, hence the humidity, by first measuring the air temperature and the wet bulb temperature (this latter is the temperature of a thermometer bulb which is surrounded by a muslin kept wet with distilled water).

Rain

Precipitation consisting of drops of water falling from layer cloud extending to considerable heights in the atmosphere. Much or all of this cloud will be at sub-zero temperatures.

Ridge

A pressure pattern of curved isobars convex towards low pressure, often an extension of a high. The area of relatively high pressure between successive lows.

Showers

Liquid or solid precipitation from convective or cumulus type cloud. Frequently intense and sudden in character a shower can stop and start abruptly while rain may persist for many hours.

Sleet

The name given to a mixture of rain and snow or to melting snow.

Storm cones

Triangular shapes hoisted apex upwards or downwards as visual warnings of winds of gale force 8 or more. The system was initiated by Admiral Fitzroy in 1860 and, for a time, included the use of cylinders to warn of severe gales.

Stratus

A generally grey uniform layer of cloud with a very uniform base and which may give drizzle. When prefixed by 'alto' or 'cirro' the cloud is very much higher.

Swell

Wave motion which is caused by winds at some distance away often persisting for some while after the winds have died down.

Temperate latitudes

Broadly speaking that part of the earth lying between the arctic and the subtropical highs.

Trade winds

Winds which persist for much of the year to the south of the subtropical highs and which blow towards the equator but with an easterly component. They are, therefore, north-easterly in the northern hemisphere and south-easterly in the southern.

Trough

A pressure pattern of curved isobars which are convex towards the higher pressures. A front will always lie through (that is, along) a trough which may or may not be very marked. A trough, however, need not be associated with a front.

Ventimeter

A simple hand-held instrument for measuring the wind speed or force.

Warm sector

The area behind the warm front and ahead of the cold.

Introduction

The Chief Instructor at the National Sailing Centre, Cowes, in his introductory speech to newcomers, used to say (and probably still does) that sailing is a high risk activity. Just how high the risk can be was tragically illustrated in the Fastnet race in 1979 when many yachts suffered knockdowns beyond the horizontal and a number of lives were lost. A storm had developed very quickly, so quickly in fact that warnings of the strong winds, although issued, were not heard by many of those competing until the winds had increased to force 9 or 10. Whether the result would have been any different had the warnings been heard is not possible to say – most of the fleet was too far from land to seek shelter. Some lives might have been saved if the crews had been better prepared for the storm; but that is a matter for speculation. Nevertheless, this incident does demonstrate the necessity for the yachtsman to be fully aware of weather hazards and to develop a working knowledge of meteorology so that he appreciates the limitations of forecasts and can determine when the weather is likely to worsen seriously.

In this book I have tried to cover enough of the basic theory – the whys and wherefores – of meteorology for the yachtsman to understand better the difficulties facing the forecaster. The descriptions of local weather effects will, I hope, lead to a realisation of just how the weather which a yachtsman experiences might differ from that described, even in a 'correct' forecast. The overall message of this book is that the forecast services provided should be used, not with blind faith but with intelligent discrimination. In this I am following the intention of the Royal Yachting Association in defining the meteorological part of the syllabus for the Yachtmaster's certificate which sets down reasonable standards of achievement for such nautical skills as navigation, pilotage and boat handling, as well as first aid and meteorology.

In writing this book I have leant heavily upon experiences gained while lecturing and running courses designed to help yachtsmen achieve the Yachtmaster standard in meteorology; I have attempted to answer many of the questions which I have been asked by sailors over many years and I have tried to point the way to the reader to show him how to learn to be wise in the ways of the weather. The skills needed cannot be acquired exclusively by reading – there must be a substantial effort and application by the yachtsmen to learn by the hard road of experience.

It will be clear by now that I have written this book with the cruising yachtsman foremost in my mind; the kind of sailor who will cruise or, perhaps, race around our shores and who might venture across the Irish Sea or down to Spain. Nevertheless, much of what I have written should provide a starting point for dinghy sailors at one end of the sailing spectrum and ocean cruising or racing yachtsmen at the other. The dinghy sailor should graduate from this book to *Wind and Sailing Boats* by Alan Watts who covers, in greater detail than I am able, such matters as gusts and other local wind abnormalities. The ocean sailor is advised to study the HMSO publication *Meteorology for Mariners* which is the definitive textbook on meteorology for seamen containing the information needed by masters and mates to gain their various certificates of competency up to and including that of Extra Master.

To return to my opening remarks, a word of warning should be sounded that, even with the deepest knowledge of meteorology, the sailor will always be at risk because of the weather. The most efficient application of all the techniques described will not eliminate risk completely any more than the highest standard of seamanship or the most careful use of highly sophisticated navigational aids will. Sailing will always be a high risk activity. Whatever precautions are taken accidents can still happen, gear fail and forecasts go wrong. The only way that a sailor can avoid sailing risks is to stay at home! To most of us that would be unthinkable and I think that to most sailors the acceptable compromise is to minimise the risks inherent in their sport by careful maintenance, good navigation and the prudent use of weather forecasts.

Finally, I have, throughout this book, referred to yachtsmen and forecasters as 'he'. This is in the interests of brevity and not through any intentional display of male chauvinism; there are, after all, many extremely competent female sailors and meteorologists and in both activities ladies participate on an equal footing with men. I have followed the lead of Eric Twiname in his book on the international racing rules in which he defines a yachtsman, helmsman and, in my case, forecaster as a person of either sex who takes part in these activities. All will be referred to as 'he'.

Good luck! Good sailing! May all your winds be fair!

1

Winds – Why We Have Them

The word 'weather' is usually taken to encompass all the various elements which describe the state of the atmosphere as it affects man and his activities. These elements are wind, temperature, visibility, cloud, precipitation (an omnibus word used by meteorologists to include rain, drizzle, sleet, snow and hail) and humidity (the water vapour content of the air). The study of the weather in its broadest sense is known as meteorology, with weather forecasting being the application of the science to the practical problem of weather prediction. As we shall see in this chapter the physical processes responsible for determining the behaviour of the atmosphere are not, in themselves, difficult to understand. What is more difficult to grasp is the complexity of the interactions between the various processes.

The desire to predict, or even influence, the weather is strong and there have always been those, such as farmers and sailors, whose livelihood and physical safety are weather-dependent. It is, perhaps, not surprising that such people living in close contact with the elements should develop the ability to observe and remember the most likely sequences of weather, either at various times of the year or following the occurrence of certain wind directions, cloud types or patterns. By careful observation man has built up a wealth of weather lore and the use of such techniques has some measure of success, especially over short periods. The professional meteorologist should not decry them, and some of the more popular sayings are discussed briefly later in this book. Nevertheless, the atmosphere is a complex mechanism and the most reliable approach to weather forecasting is the traditional scientific one of attempting, first, to understand the true nature of the problem and then to apply that knowledge in a logical manner. The product of the application of meteorological knowledge to the forecasting problem is made available to the general public through a wide range of forecasting services.

The yachtsman is only one of the many categories of users of the national weather forecast services but is in a rather vulnerable position compared with the majority of people, in that his personal safety and that of his craft can be put at extreme risk by the weather.

Of all the weather phenomena the two of paramount importance to the yachtsman are the wind and the visibility. The former always seems to be discussed the most, in both textbooks and in forecasts, probably because of the dramatic effects resulting from gales and storms. It is, nevertheless, visibility with which the sailor is often most concerned. Given the choice many sailors would probably opt for being caught in mid-Channel in a force 9 rather than in thick fog and a light wind. However, it is the wind which is of the greater importance to the meteorologist because it is the wind systems which determine the movement of the air masses with their different characteristics – including visibility – and we shall, therefore, follow the traditional pattern in considering the wind first.

Wind systems

There are three types of wind system which can be observed. Firstly, well known to sailors for very many centuries, there are the winds which persist for long periods of the year in some areas of the world or which set in with great regularity and last for months; these are, respectively, the *trade winds* of both hemispheres and the *monsoons* (from the Arabic 'mausim' meaning season). Monsoons are most marked in the northern hemisphere but also occur, although only weakly, in the southern. Secondly, there are the *mobile wind circulations* which develop through well recognised life cycles as they move and which are a dominant feature of the temperate latitudes in which the United Kingdom is placed. Thirdly, there are various *locally generated winds*; some of these are topographically induced and often are given local names, such as the 'Bora' and the 'Mistral'. Other local topographical effects are sea breezes, eddies around headlands and other distortions in the flow of air that we call wind. A different category of locally produced winds are the short-lived gusts which result from convection currents, heavy showers or, simply, as a result of turbulence when the air flows over rough ground.

The climatological pages of a geography atlas show the trade winds. These blow north-easterly in the northern hemisphere and south-easterly in the southern hemisphere meeting near the equator in the 'doldrums', an area of light and variable winds with heavy showers, often accompanied by thunder. The doldrums move northwards and southwards during the northern hemisphere summer and winter respectively following the sun and being associated with the hottest parts of the surface of the earth. Over the land, which heats up quickly, this area is a little to the equator side of the sun when at its maximum (noon) elevation; over the sea the warmest water is never very far from the equator having a maximum latitude of about five degrees north or south. This is because of the greater heat capacity of the seas relative to the land and their consequential slower response to heating from the sun in the summer. When the sun moves from the southern to northern hemisphere the south-east trade winds of the southern hemisphere are drawn across the equator to meet the north-east trades in the doldrums. On crossing the equator the south-east winds turn to become south-west. Conversely the north-east trade wind, crossing the equator in the southern summer, turns to the north-west.

To the poleward side of the trade wind belts are some more areas of light and variable winds known as the 'Horse latitudes'. It was in these latitudes, 30–40 degrees north, that sailing ships often became becalmed and, running out of water and fodder, used to jettison horses which were being transported to the Americas.

Less important to the sailor but not so to the meteorologist are the winds blowing away from the polar regions; not usually very strong, these normally have an easterly component. In this chapter we shall try to understand the reasons for some of these winds and other important features of the atmosphere.

Air pressure

Perhaps the first question of all to answer is the fundamental one of why there is air movement at all. But first it is necessary to know the meaning and significance of air pressure. When we measure air pressure with a barometer we are, in effect, measuring the force being exerted by the mass of air above the barometer; in other words we are measuring the weight of the atmosphere at that point. This is best demonstrated by means of the simple barometer of Torricelli. This is a tube rather less than one metre long, completely filled with mercury, a stopper inserted in the open end, turned upside down and the end put into a reservoir of mercury with the stopper then removed. The force which keeps the mercury up the tube is the force of the atmosphere acting downwards, as shown in Fig. 1. If we took the

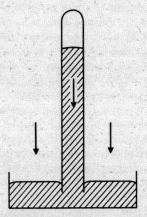

Fig. 1 A simple mercury barometer in which the weight of the atmosphere pressing down on the mercury in the reservoir is balanced by the weight of the mercury in the tube.

barometer up a hill or upstairs in a tall building we would see that the atmosphere would then support a shorter column of mercury. That this should be so is reasonably obvious because there would be a smaller mass of air above the barometer and, therefore, a lesser weight pressing down on the mercury in the reservoir. The decrease in weight of the air is simply the weight of the column of air through which we would have passed in ascending the hill or climbing the stairs. We can therefore conclude that if different pressures are measured at two places on the surface of the earth then there must be, literally, a greater mass of air over the place with the higher barometer reading.

Household barometers, and many of those used for engineering or laboratory work, are calibrated in terms of the length of the mercury column which would be supported by the atmosphere; pressure values are often quoted in such cases of about 30 inches or 76 centimetres. The meteorologist, however, uses the unit of pressure known as the *bar*. One bar is approximately one standard atmosphere and is divided into *millibars* so that typical pressures that you hear quoted, in the shipping forecast for example, range from about 960 to 1030 millibars with extreme values in the area of the United Kingdom of about 920 and 1060 millibars.

(In the usual scientific units 1 bar is equal to 1,000,000 dynes per sq cm in the cgs system or 100,000 newtons per sq metre in the modern SI units. One newton per sq metre is nowadays known as one pascal, so that a sea level pressure of 1 bar or 1000 millibars can also be written as 100,000 pascals or 100 kilo pascals. For normal meteorological purposes the millibar (mb) is the most convenient unit and this is the only one with which you need be concerned.)

Weather charts that you see in newspapers or on television have lines of equal pressure drawn on them known as *isobars*. Isobars are, in effect, a representation of the distribution of the mass of air over the surface of the earth in precisely the same way that the contours of an Ordnance Survey map represent the height of ground above sea level. One very significant difference between the two types of map is that isobaric charts present an ever-changing picture of the atmosphere while geographic contours barely change in a millennium. In a similar fashion to the geographic charts the choice of intervals between the values of consecutive isobars is a matter of convenience and, to a large extent, depends upon the scale of chart being used and the degree of detail required. The United Kingdom Meteorological Office uses 4 mb intervals on many of its charts – 992, 996, 1000, 1004, etc. On large charts of small areas intermediate isobars at 2 mb or even 1 mb might be drawn while on charts covering very large areas intervals of 8 or even 16 mb might be more appropriate.

To return to the question of the air and its movement, it is a fundamental law of physics that if there is a high value of some quantity in one place and a lower value nearby then there will be a tendency for a levelling out process towards achieving the same value everywhere. One example of this is a metal bar heated at one end and cooled at the other. Heat will flow along the bar from the hot end to the cold. Another example, which demonstrates better what the atmosphere is trying to do, is a bowl of syrup from which a spoonful of syrup has been removed creating a situation where there is more syrup around the outside of the bowl than in the middle. The effect is that the syrup moves from the outside to the middle. The deeper the hole then the faster the rate of flow of the syrup. The steeper the sides of the hole the faster the movement of the syrup and the less steep the sides the slower the movement of the syrup. The driving force for the syrup obviously depends upon the difference in height of the syrup between the outside of the bowl and the middle. If rice grains were mixed into the syrup then the actual movement within the syrup would be seen. There would be downwards movement of the syrup around the sides of the bowl where the height of the syrup was greatest and upwards movement in the middle as the hollow filled up. Linking the two areas of vertical motion there would be horizontal movement from the outside towards the middle across the bottom of the bowl but the speed of the horizontal movement would decrease with increasing height in the syrup. There would be no noticeable horizontal movement across the surface of the syrup.

If we now think about our two places on the surface of the earth where there are two different pressures or, as we have seen, two different amounts of air as measured by our barometer, then there will be a driving force, precisely in the same way as for the bowl of syrup, trying to make the air move across the surface of the earth from the place with the higher pressure towards the lower pressure. If we have a region where the pressure is lower in the middle, then the air from the surrounding areas will try to fill up the hole; similarly, if we have a region of high pressure surrounded by a lower one then the air will try to flow from the 'hill' to the nearby 'valleys'.

But does this really happen?

Look at the isobars on the television weather map and then look at the wind arrows on the caption chart, or look at the weather map in the newspapers and the accompanying forecast. Do the winds blow across the isobars from high to low pressure as would be expected from the analogue of the bowl of syrup? The answer, very clearly, is NO! The wind seems to blow, more or less, *along* the isobars.

What about the wind speed?

If you look at the strengths of the wind on the television chart or in the newspaper forecast then you will see that the strongest winds are in the regions of greatest slope of pressure (the greatest pressure gradient), that is where the isobars are the closest together, in much the same way that the closest contours on the Ordnance Survey map indicate the areas of steepest hills. Part of the syrup analogue seems to explain the wind and its behaviour – the steeper the pressure gradient the stronger the wind – but what about the direction?

The reason that the wind does not flow across the isobars is simply that we are on a rotating earth. The effect can be demonstrated by going to a children's playground where there is a roundabout – preferably with a child! Start the roundabout spinning in an anti-clockwise direction and stand on the roundabout near the rim with the child in the centre. Now throw a ball gently to the child. Observe the flight of the ball. Then ask the child to throw the ball to you. Again, what happens? Now both stand near the rim about a quarter the way round from each other and repeat the experiment. Then reverse the direction of rotation of the roundabout and repeat the whole sequence. Fig. 2 shows the path of the ball in each case.

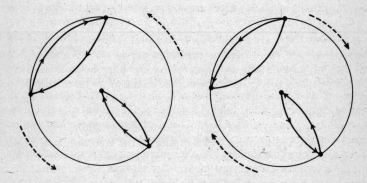

Fig. 2 Paths of a ball thrown between two people on a turntable spinning in clockwise and anti-clockwise directions.

Anything which moves from one place on a spinning surface to another will appear to an observer on the spinning surface to be deflected. The deflection will be to the right for anti-clockwise motion and to the left for clockwise motion, and will occur whether it is a ball being thrown across a roundabout, a long distance shell, a rocket or the air moving across the earth. This result is known as the *Coriolis effect*, after the Italian mathematician who discovered it. The result of the Coriolis effect is that as the air starts to move across the isobars, because of a force which depends upon the isobar spacing, the spin of the earth causes deflection to the right in the northern hemisphere where the earth is spinning in an anti-clockwise sense when viewed down from over the north pole. In the southern hemisphere the deflection is to the left because, here, the earth is spinning clockwise when viewed from over the south pole.

The Dutch professor, Buys-Ballot, was the first to describe this result in 1857, and the law to which he gave his name states: 'If in the northern hemisphere you stand with your back to the wind the pressure will be lower on your left hand than your right. The reverse is true in the southern hemisphere.'

The reversal from the northern to southern hemisphere rule, rather obviously, must occur at the equator and, here, the wind does blow more or less across the isobars from high to low pressure. Very quickly on either side of the equator the spin of the earth starts to take effect and by about 10–15 degrees north and south the winds are seen to blow nearly along the isobars again. Away from the equator the air moves around rather than into areas of low or high pressure so that lows and highs retain their identity for long periods. Near the equator, on the other hand, because the air moves directly from high to low pressure, the lows quickly fill up and the highs flatten out.

It should, perhaps, be made clear that the full deflection because of the spin of the earth does not occur as soon as the air starts to move. If a pressure gradient were suddenly created then the initial movement of air would be direct from high to low but then the Coriolis effect would begin to be noticeable and, before long, the air would settle down to a direction along the isobars. The scale of movement to be affected by Coriolis has to be at least a few miles. Coriolis does not determine the way in which the water goes down the bath plug hole: that is a function of the shape of the bath; nor does it account for the way that a fast bowler in cricket can make the ball swing into the right handed batsman: that is a question of the aerodynamics of the cricket ball.

For reasons which will be discussed in Chapter 4 the relationship between wind and isobar spacing, although not precisely one to one, is, nevertheless, sufficiently good for many practical purposes. For example, given sufficient pressure values, it is possible to draw isobars and so deduce the wind pattern and, similarly, with relatively few pressures but with the addition of enough wind observations to determine the wind pattern it is possible to draw with reasonable confidence a coherent set of isobars. From these isobars wind speeds and directions can be deduced in places where there are no wind reports. In particular, if a chart of a forecast of an isobar pattern is produced then it is possible to deduce the forecast wind field. Forecasters use special scales to enable them to relate isobaric patterns to winds.

● 1008

● 1008

● 1004

● 1012

Fig. 3   Values of barometric pressure to which isobars could be drawn in countless ways.

Fig. 3 shows some values of pressure with no winds and by trying for yourself you can see that it is possible to draw different patterns of isobars to these pressures. There would have to be several more pressure readings before the isobars could be drawn with any confidence. Now add some winds and use a scale relating pressure gradient to wind as in Fig. 4a; this scale is called a *geostrophic scale*

Fig. 4a   A geostrophic scale. Used with an appropriate scale of chart and laid across patterns of isobars with the left hand end of the scale on one isobar the point at which the next isobar crosses the scale indicates the wind speed. This particular scale is used with isobars at four millibar spacing in Fig. 4b–4d.

and it gives wind speeds for various isobar spacings for straight isobars.

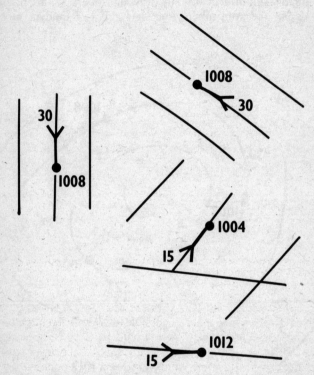

Fig. 4b Wind arrows have now added to the pressures. Using the geostrophic scale spacings of the isobars near each pressure value have been deduced and short lengths of straight isobars drawn along the wind directions shown by the arrows.

Fig. 4b–4d shows the various stages in drawing isobars to winds and pressures.

Remembering that a pressure and a wind allow you to see not only the direction of the isobars at that point but also the spacing, you can first draw parts of isobars near each pressure value. Then, by extrapolation, the lines can be joined up right around what must be a low centre, assuming that this is in the northern hemisphere. Finally, by continuing the same spacing of the isobars into the centre of the low, a central pressure can be estimated. The finished product should be reasonably smooth because, by drawing isobars, we are attempting to depict the atmosphere for what it is – a fluid in motion.

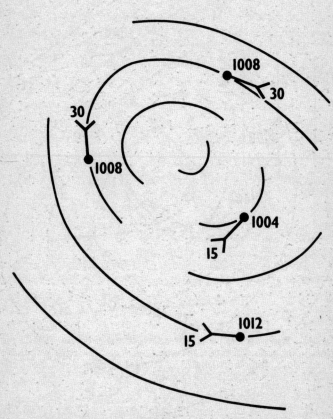

Fig. 4c Some of the isobars have been extended and joined up. In order to follow the wind arrows the isobars have had to be curved. Precisely how gradual or how sharp should be the curve is a matter for conjecture.

Sometimes there are sharp bends in the isobars, as we shall see but, for the most part, the winds, and therefore the isobars, flow smoothly. The act of extrapolation both into and away from the centre of a pressure system, be it high or low, requires a certain amount of faith and experience.

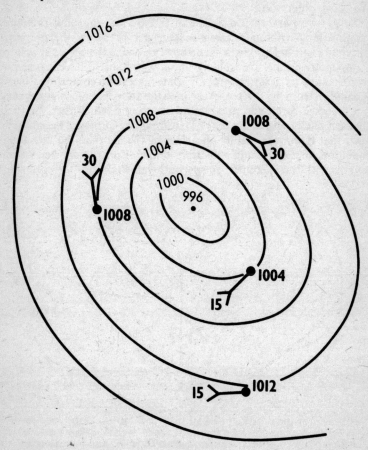

Fig. 4d Further extrapolation of the isobars produces a complete circulation from which both winds and pressure can be estimated over the whole area. In practice, of course, the pressures will not all be whole, even numbers and the positions of the isobars will have to be interpolated between the pressure values.

Because of centrifugal effects the winds around a low pressure centre will be less than the wind for the same isobar spacing with straight isobars; around a high the winds will be stronger than the straight isobar wind with the same spacing. The stronger the curvature of the isobars the greater the difference of the wind from the geostrophic value.

So far we have rushed ahead to talk about wind and pressure relationships without pausing to consider just how the pressure differences arise. Unlike the syrup bowl analogue there is no spoon to scoop a hole in the atmosphere. Initially, pressure differences result from heating of the atmosphere in some places and cooling in others. Radiation from the sun passes through the atmosphere, wherever it is not reflected by clouds, and warms the ground which then radiates the heat back towards space. The nature of the outgoing radiation is such that it is more readily absorbed by the air than the incoming radiation. The incoming radiation is mainly in the shorter wavelengths with the maximum energy in the green part of the spectrum

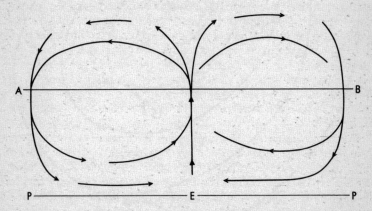

Fig. 5   Air circulations in a room (or over a stationary earth) which is heated in the middle (or equator) E and cooled at the ends (or poles) P. Air ascends over E and descends over P. Below the level AB air is moving equatorwards and pressure at ground level must therefore be greater at P than at E. Above level AB the air is moving polewards and, at these upper levels, the pressure must be higher over E than over P. Level AB is a crossover level with no horizontal air movement and, therefore, the pressure at A and B will equal that at the same height over E.

while the outgoing radiation is largely in the infra-red. The air is thus heated more where the underlying ground is hot, and therefore radiating more heat, than where the surface is cold.

If the earth was uniformly covered with land or sea and was not rotating then we would have two simple convection cells similar to a room with a hot radiator in the middle and a cold window at each end, as in Fig. 5. The wind which would blow from the direction of the poles at low level implies that the pressure near the equator at the surface of the earth would be low while over the poles the pressure would be high; this follows from the syrup analogy. The reverse flow at high level is needed for continuity but is, in any case, consistent with the pressure exerted by the air above some level such as AB in Fig. 5. Because the cold air over the poles is denser than that over the equator a column of air of, say, 5000 metres depth will have a greater mass than a similar depth of air over the equator. As a consequence, if we start with a higher pressure over the poles than the equator then at some height, because of the greater mass of air below that height in the cold air than in the warm, the pressures will be equal. In other words, at some height there will be just as much air, in terms of mass, above that height over the pole as above that same height over the equator. At that level there would therefore be no tendency for air to move in either direction. Further up the same effect of pressure decreasing more rapidly with increasing height in the cold air over the pole than in the warmer air over the equator would lead to the pressure being lower over the pole than over the equator. This is consistent with the high level part of the convection circulations in Fig. 5.

Keeping, for the time being, with a uniform earth but now allowing rotation has the effect shown in Fig. 6. The direct cell, starting with ascent over the equator, now has its corresponding descent at about 35 degrees latitude. There is some descending air in high latitudes but at a slower rate than would be the case in the hypothetical situation of a stationary earth. The low pressure area near the equator is still present but, just as descent over the pole implied high pressure there, the descent around 35 degrees north and south implies high pressure areas there also. These high pressure areas are shown on the climatological atlas as a 'family' of highs right around the globe with the Azores high being one member. The air moving equatorwards at sea level from these highs is deflected eastwards by the Coriolis effect – to the right in the northern hemisphere and to the left in the

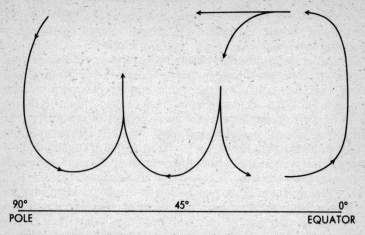

Fig. 6   A simplified vertical cross-section through the atmosphere above a rotating earth which is heated at the equator and cooled at the poles (one hemisphere only is shown).

southern – to give the north-east and south-east trade winds of the two hemispheres. The very light and variable winds in the middle of the high belt between 30 and 40 degrees (the subtropical high pressure belt) are the light winds of the Horse latitudes mentioned earlier.

The stability of the large-scale mechanism of the direct heat-driven cell, described by Hadley in the eighteenth century and named after him, is the reason for the persistence of the trade winds. To the poleward side of the subtropical highs the air moves away from the highs to the north (in the northern hemisphere) to give a mainly south-westerly flow of winds (north-westerly in the southern hemisphere). Air moving away from the poles is also deflected by Coriolis to give north-easterly or south-easterly winds near the north and south poles respectively.

Fig. 7 shows the low level wind patterns described so far with air reaching mid-latitudes originating from either polar regions or the subtropical highs. So far we have only allowed a uniform earth and comparison of Fig. 7 with the average flow patterns of the real atmosphere, as shown in the climatological pages of the geography atlas, will show some marked differences. There are also differences in the real atmosphere comparing summer and winter. For example,

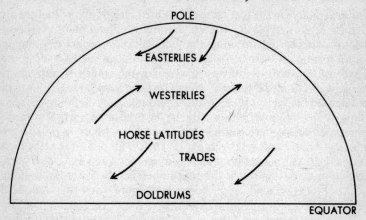

Fig. 7 Idealised wind patterns at the surface of the earth resulting from the motions shown in Fig. 6.

the subtropical highs occur over both land and sea in the winter but only over the sea in the summer. Superimposed on the differential heating between poles and equator there are also very significant differences in the heating of land and sea around the same latitude band.

Ignoring this complication for the time being we will return to the direct Hadley cell between the equator and the subtropical high pressure systems. The air of the high pressure belt is descending and is, as a consequence, dry while the ascending air in the equatorial belt is wet. To understand the reasons for this we must remember that air can hold water as an invisible vapour and that the amount of vapour which can be held increases with increasing air temperature. This can be seen in a warm room if the curtains are drawn back when the air is cold outside. The air inside the room near the windows is cooled by being in contact with the cold glass and condensation occurs on the windows. The air has been cooled to such a temperature that it cannot hold as much water vapour as it did previously at the higher temperature. The mechanism by which the air jettisons the surplus water vapour is known as *condensation* and the temperature to which the air has to be cooled for condensation to take place is called the *dew-point*, being the temperature at which dew can form.

Air in the tropics is very warm and there is a plentiful supply of water available in the seas to be evaporated into the air so that the air in these regions, typically, has a high water vapour content. As the air rises it cools by expansion so that condensation occurs resulting in cloud formation. Further ascent gives continued cooling and more condensation until the drops of water in the clouds become large enough to fall in the form of rain. There are very large amounts of condensed water in these tropical clouds and the rain is, consequentially, often very heavy. By the time the air is starting to move polewards at high levels it has lost much of its water vapour to the rain and has become relatively dry. When the air descends in the high pressure areas it warms up again by compression and any cloud starts to evaporate so that by the time the air reaches the lower levels of the atmosphere it is capable of holding much more water than it does in fact contain.

The typical weather of these subtropical high pressure areas is dry with little or no cloud over land. There may be some cloud in the lowest levels over the sea where evaporation has occurred so allowing the formation of cloud because of ascent of air by means of some very limited convection or, simply, by means of turbulence. It is not surprising that the major deserts of the world are associated with this high pressure belt. Similarly, over the poles the air is relatively dry because the low temperature of the air means that the water vapour content must be low whatever the vertical motions.

In the trade winds the air moving equatorwards is subjected to strong heating from below resulting in vigorous convection. Over the oceans this means that the water vapour from the sea is pumped to great heights in the atmosphere. The convective effect is enhanced greatly in the regions where the trade winds from the two hemispheres meet. This area of convergence of air from the north and south of the equator is called the *Inter-Tropical Convergence Zone* (ITCZ) and is marked by large convective clouds, cumulus and cumulo-nimbus often with thunderstorms. The latitudinal position of the ITCZ varies throughout the year, following the sun as it crosses the equator but usually lagging a little behind and having its position over the warmest part of the surface of the earth.

The ITCZ sometimes produces particularly vigorous areas of convection which can be observed to move westwards (because of the easterly component of the winds in these latitudes). The very strong ascent in the convection causes an inflow at low levels in the atmosphere and this inflow of air is deflected by means of the Coriolis effect to produce a circulation of winds, anti-clockwise in the northern hemisphere. Continued convection and the effect of the circulation so formed combine to develop into a very vigorous tropical storm which may be sufficiently active to be called a hurricane if it is over the Atlantic or eastern Pacific.(Over the China Sea such a storm would be called a typhoon.) These storms continue to move westwards but, being some little way from the equator, are subject, as entities, to the deflection of Coriolis and are themselves deflected to the right (in the northern hemisphere). Those hurricanes in the southern part of the north Atlantic continue turning until they have turned completely on to an easterly track. By this time they are acquiring the characteristics of the more usual temperate latitude low pressure systems which we shall be discussing later.

The hurricane or typhoon gains much of its energy from the sea by means of the heat released during the condensation process so that if the storm passes over land, as sometimes happens with devastating consequences over the southern states of the USA for example, then the supply of water vapour, and therefore energy, is cut off and the storm quickly loses its vigour. While the hurricane remains over the warm sea it is a self-generating mechanism. (We shall return to this self-generating aspect of weather systems when we look at the low pressure areas which affect the British Isles.) The role of the sea as the root cause of the hurricane is dependent upon the sea temperature to such an extent that hurricanes have never been observed over the Atlantic to the south of the equator although they are a regular occurrence during the late summer and early autumn in the northern hemisphere. Apparently, a sea temperature of at least 28°C is required for hurricane formation and, while this value is reached in the northern hemisphere, the highest southern Atlantic sea temperature is about 27°C.

The monsoons, another example of very large-scale wind systems, are the result of the modification of the trade wind circulation either by the very hot ground of India, Arabia and south-east Asia in the northern hemisphere summer or by the very cold ground of much of Eurasia in the northern hemisphere winter. In the winter the very strong outflow from the Siberian high gives a much enhanced north-east trade wind over the Indian Ocean. In the summer the high disappears and is replaced by a deep and extensive low pressure area over south-east Asia. The inflow to this low is so strong that, instead of the north-east trade winds which would otherwise blow over the Indian Ocean, there are strong south-westerly winds. The southern hemisphere south-east trade winds cross the equator, turn to the south-west in so doing and are then, in effect, simply sucked up into the monsoon low pressure over northern India. The very moist air of the south-west monsoon, after crossing the Indian Ocean, is then forced upwards over the Himalayan foothills and so gives extensive rain. There are also lesser monsoon effects associated with the heating over North Africa and South America as well as Australia. These are, however, relatively feeble affairs compared with the Indian monsoon.

2

Air Masses, Fronts and Weather Systems

The previous chapter described the very large-scale effects arising from a rotating earth heated at the equator and cooled at the poles. In particular, it was shown that the air reaching mid-latitudes originated either from the polar regions or from one of the systems of the sub-tropical high pressure belt. We will now look at the characteristics of these two types of air mass and the weather associated with them, and then see what happens when the two air masses meet.

Air from the polar regions

Air from the polar regions is, more or less by definition, cold throughout most of its depth and fairly dry. As such air moves southwards, as it must do in order to reach the British Isles, it is warmed at low levels by contact with the sea. The result is convection which takes place in the form of bubbles rising randomly, rather like the bubbles of air from the bottom of a heated saucepan. The air rising is more moist than it was initially because by this time there has been evaporation from the sea surface. The bubbles of air continue to rise while they have buoyancy, which came initially from being warmed more than the neighbouring air. During the ascent, however, cooling takes place because of expansion resulting from the pressure decrease as the bubble gets higher in the atmosphere. If sufficient cooling takes place then the bubble ceases to have any buoyancy and it rises no further. The height at which this happens depends on the rate at which the temperature decreases with increasing height in the atmosphere through which the bubbles are rising, as compared to the rate at which the temperature falls in the bubbles themselves, as they ascend.

If the temperature decrease in the surrounding air is such that the bubbles can rise quickly and to great heights then the air is said to be unstable. Air which is coming directly from polar regions is very unstable. In air which comes towards the British Isles on a roundabout route across the Atlantic (the returning polar maritime air masses beloved of geography textbooks) then there will have been some warming of the air by the sea and it will be less unstable.

During the rising and cooling process the air may cool to the dewpoint temperature so that, with further cooling, condensation takes place on small nuclei – mainly salt particles of which there are usually many in the air, especially over the sea – to give clouds. During the condensation process heat, known as *latent heat*, is released and increases the buoyancy of the rising bubbles so that they rise higher. Although further cooling may take the temperature of the air below 0°C the drops may well remain liquid – very small drops can be 'supercooled', occasionally as low as −40°C and frequently as low as −10° to −20°C. If the drops freeze then more latent heat is given out and the bubbles receive another burst of energy and so can continue to rise.

Cumulus cloud

The type of cloud formed by this convection process, aided by the release of latent heat, first of condensation and then of fusion, is known as *cumulus*, from the Latin word meaning heap, see Plate 1. Sizes of cumulus clouds can vary from a depth of 1000 ft or less to 20,000–30,000 ft or so in polar air masses and to 40,000–50,000 ft or more when they form in association with the ITCZ, described in the last chapter. The size of the clouds depends upon the buoyancy of the air initially, the amount of heat released during the condensation and freezing processes and the number of bubbles which combine to form one cloud. The last reason is because the water drops in the rising bubbles evaporate rapidly at the edge of the cloud and in so doing reduce buoyancy (for the reverse reason that condensation increases buoyancy). If, however, one bubble is protected by the very moist air of other bubbles then it may be able to rise even though the others surrounding it have lost their buoyancy. The combination of a large number of bubbles seems to have the effect that a cumulus cloud is about as large horizontally as it is vertically.

Droplets of water in clouds grow partly by the continued condensation of water vapour and partly by means of collision with each other. If droplets grow large enough then they can fall faster than the air is rising and, hence, eventually fall out of the cloud as rain. If the air in the cloud is rising vigorously then drops can become quite large before falling from the cloud. Shallow cumulus clouds are associated with weak upcurrents and only give light showers or showers which do not reach the ground (being only small the raindrops evaporate in the air below the cloud) or, perhaps, no showers at all. Deep cumulus clouds will be formed by strong upcurrents and, therefore, are quite likely to give heavy showers which may well be accompanied by thunder. The drops in such large shower clouds may freeze rapidly and become hail. These very large convective clouds are known as *cumulo-nimbus* because, when seen from below, they are very dense and black in appearance, nimbus being the Latin word for black.

The final burst of energy needed to create the conditions for the thunder and hail seems in some way, not yet fully understood, to be associated with the freezing process of the supercooled drops of water high up in the cloud. The typical shape of the top of the thunder cloud is often an anvil shape composed of ice particles. The ice particles evaporate very slowly, and, being very small, only fall slowly so that they get carried away by the strong winds high up in the atmosphere. The main body of the cloud, being in lighter winds, moves less quickly than the top and so the top of the cloud appears to spread away from the bottom. The top of the cumulo-nimbus cloud thus has a different appearance from the lower parts of the cloud which still have the bubbly look of convection cloud; Plate 2 is a good example.

The vigour of showers in air of polar origin depends mainly on just how cold the air is and how warm, by contrast, the sea is. Air coming fairly directly from the poles will be very cold and unstable and will thus give heavy showers. Air with a long sea track, will have been warmed and so will not be so markedly unstable and may give less heavy showers although there is some compensation in that this air, having had a long sea track, will have a higher water content and so the difference in shower intensity might not be as great as might be expected from consideration of the instability of the air alone.

Polar lows

Particularly in winter there can be very vigorous convection in polar airstreams leading to an inflow of air, in a similar fashion to the early stages of the hurricane and, because of the Coriolis effect, the formation of a small vortex. These *polar lows*, as they are called, are feeble affairs in relation to their tropical counterparts because the air, being colder, can hold less water vapour and there is, thus, far less heat energy available to be released by condensation. Although they do not give major wind systems they do, nevertheless, create some quite severe weather with heavy snow falls and are notoriously difficult to forecast partly, at least, because they are often difficult to detect in their early stages. Polar lows may also form as eddies in a northerly airstream flowing around Iceland and Greenland. The eddies break away from the land and then are carried with the general flow of wind southwards; convection within the eddies is enhanced by the inflow associated with the vortex and the extra inflow so created helps to maintain or increase the polar low circulation.

Visibility

One of the most important characteristics of polar air masses is the visibility which is almost invariably very good as a result of the deep convection causing any pollution, whether man made or not, to be mixed throughout the atmosphere and so become well diluted. Also, the effect of showers occurring in polar air will be to wash dust and other particles out of the air and so help to give good visibility. The visibility, in fact, will only be poor in the showers themselves – in the really heavy showers or when the showers are of snow, the visibility will be very poor.

Convection

Convection will occur over the sea at all times of the year in polar air masses but may be more vigorous in the winter than the summer because of the very cold air and the relatively warm sea. Over the land convection will not occur during the cold days of winter nor at night at any time of the year; the ground simply is not warm enough to give the air the buoyancy required. At times of the year when the sun rises sufficiently high in the sky to heat the land then convection will occur diurnally in these polar air masses. At any time of the year or of the day, however, an on-shore wind will carry cumulus clouds, which have formed over the sea, on to the land and some little distance inland although they decay rapidly and, unless they are very big clouds and the wind is fairly strong, they do not penetrate to any great distance from the coast. It is this mechanism on a winter's day which can give showers in a northerly airstream over those parts of the British Isles exposed to that direction. A small swing to the east in the wind and the showers all fall over east coast districts while a slight shift of the wind to a point west of north leaves the east coast dry and gives the west coast a showery day. A very slight error in the wind forecast can thus give a very noticeable error in the weather forecast and one which can be glaringly obvious subsequently when the weather is cold enough to give snow.

Air from the subtropical highs

Turning our attention away from the poles to the area of the subtropical highs we note that air in these highs will be warm and, because of the descending motions, dry. As this air moves northwards the underlying sea temperature will usually decrease and the air at low level will be cooled by contact with the sea. There will, consequently, be no tendency for convection. Any vertical movement in the lowest layers will be as a result of turbulence arising from friction between the air and the surface of the earth. The air will acquire water vapour by means of evaporation at the sea surface and the turbulence will distribute the water vapour throughout the lowest 1000–2000 ft of the atmosphere. The air that is forced to rise by the turbulence will cool in so doing and, if enough cooling has taken place, will form cloud. This cloud will be a flat, rather uniform, layer with little feature. Sometimes there will be rolls in the cloud as evidence of its turbulent origin and there may be some drizzle. If the air has been subjected to sufficient cooling and there is a high enough water vapour content then the cloud may form very low down to give sea fog. We will be discussing sea fog in more detail later but will just note in passing that, unlike the typical fog conditions over land when radiative cooling is the cause with very light winds, sea fog can occur with quite strong winds. The combination of strong wind and poor visibility is extremely dangerous to the yachtsman.

Where air of subtropical origin blows onshore then the forced ascent as the air rises over cliffs and hills can also cause cooling sufficient for cloud to form. Such cloud can persist for as long as the wind is onshore providing that there is no change in the origin of the air reaching the coast. If the air is very moist then only a very slight amount of lifting is needed to give cloud or fog. Both sea and coastal fog, which are typical of the subtropical air masses, can develop at any time of the year but they are more likely when the sea around the British Isles is still cold after the winter but the sea temperature in low latitudes is starting to increase and the air approaching from the south or south-west is becoming warmer and is therefore capable of holding more water vapour.

Even when it does not produce fog, air from the subtropics can give rather poor visibility because of the lack of any tendency for mixing to occur in depth, there being virtually no convection. Any pollution will remain trapped in the lowest layers of the atmosphere. A flight in an aircraft on such a day will show a very marked transition often at about 2000–3000 ft from very hazy, murky conditions near the ground to much clearer, very obviously clean air above.

The foehn effect

Over land, where there is not a source of water vapour, and when day-time heating from the sun is sufficient for the low cloud to evaporate, the weather associated with air masses from the south is usually fine, dry and warm. In winter, on the other hand, the cold ground and the low angle of the sun do not provide any mechanism for the cloud to break and the weather can remain dull all day. If such an air mass has to cross a range of hills then the *foehn effect* can come into play. This is the process by which the rising air on the windward side of the hills forms cloud thick enough to give rain or drizzle but on the lee side the descending air is warmed and, having lost some of its water in the precipitation, the cloud breaks. It is this effect which can give sunny and mild days on the east side of the Pennines in winter when the wind is in the west or south-west.

Thunderstorms

During the latter part of the summer or the early autumn subtropical air may reach the British Isles from a southerly direction, having passed over France or Spain. The very hot ground in these countries at this time of the year can be sufficient to give convective cloud even in this normally very stable air. This sometimes occurs over the south of England on a sufficiently hot day and the thunderstorms which result are the end to many a hot and humid day in late summer. When the storms develop over northern France they can drift northwards to reach the south coast of England during the night or early on the next day, depending on the wind speeds at cloud height. As it sometimes seems to need two or three days of such weather before the ground becomes sufficiently hot to give enough convection to create these summer thunderstorms we have the effect known to some forecasters as the typical English summer – 'two fine days and a thunderstorm'!

These summer thunderstorms, or groups of thunderstorms as they often are, cause an inflow of air at low levels. Like the polar lows the inflow motions are deflected by Coriolis and a weak low pressure, cyclonic flow can develop.

Precipitation

In this chapter the words 'rain', 'drizzle' and 'showers' have been used and perhaps it is worth explaining that the meteorologist uses these words very carefully in order to describe different types of precipitation.

Showers is the word used to describe the precipitation which forms in *convective cloud*. By the very nature of its formation it is often a sharp rather sudden phenomenon, which can pass over as quickly as it starts, possibly with blue skies before and after. Showers may be of hail from the large cumulo-nimbus regardless of how warm the air is at low levels near the ground – the hail falls so quickly that it cannot melt in the warm air. If the low level air is cold then showers may be of snow.

Drizzle is the name given to the steady precipitation composed of very small drops of an extremely uniform size; sometimes a fog will be very 'wet' and, in a sense, that is virtually drizzle. The cloud in which drizzle forms is always shallow and usually has a temperature mostly or entirely above $0°C$; it is warm cloud in which the drops are all liquid water.

Rain is the name given to precipitation which forms in cloud which is of considerable vertical extent and is not convective in origin; it is formed in deep layers of cloud much or all of which may be at sub-zero temperatures. Rain drops are larger than those in drizzle and rain may well be of a non-uniform intensity. Rain, in fact, usually forms originally as snow and melts below or just above cloud base if it is warm enough. If the temperatures are low enough below the cloud then the snow reaches the ground. It is the nature of the snow process which accounts for the very uneven drop sizes in rain.

The three types of precipitation can occur in combination; for example, cloud at low level giving drizzle may have cloud higher up giving rain so rain and drizzle may be reported together. A cloud system giving rain may have convective cells embedded in it giving bursts of showery precipitation.

Sleet is rain and snow or melting snow.

To summarise this chapter so far, there are basically two air masses reaching the British Isles and the two have quite different characteristics. Polar air is, typically, clear with good visibility, cumulus-type, convective cloud and showers. Subtropical air is generally associated with a poorish visibility, quite often with sea fog or coastal fog and a layer of featureless low cloud, with drizzle, if the cloud is sufficiently thick. Over land in the summer the weather with this air mass may often be very hot, even to the extent that there can be thundery outbreaks.

Fronts

That these two air masses meet in a well defined zone was an important discovery of a group of Norwegian meterologists during the First World War. The concept of this zone being a battleground between the two very different types of air led to the use of the word *front*, and the main boundary between these two air masses is known as the *polar front*. A reasonably careful study of observations on weather charts shows the polar front to be a quasi-permanent feature of mid-latitudes marked by a band of cloud through a considerable depth of the atmosphere and a high level belt of strong westerly winds known as the *jet stream*. The name *jet* is not given to the winds at this level because it is the level at which jet aircraft fly (although they often do fly at these heights) but because the belt of strongest winds forms a very narrow and often quite shallow zone, like a jet of water issuing from a hosepipe. The cloud of the polar front is often layered through a considerable depth of the atmosphere and is caused by the warmer, less dense air of the subtropics being forced to rise over the cooler polar air. This air mass boundary meanders to and fro rather like the boundary between the waters of the Blue and White Niles when they first meet. These meanderings are, in fact, associated with minor oscillations in the jet stream. Because of the general westerly flow of the jet these oscillations move from west to east and, hence, so do the associated areas of cloud observed on our weather charts.

Low pressure areas

Sometimes these ripples or waves on the front simply run along the front as very minor insignificant features and disappear. Those that are sufficiently large to be identified on weather charts more often grow and become the travelling low pressure systems responsible for much of our rain and bad weather. Observations in the very early stage of the life of these waves shows that these are weak, low pressure areas. Air from around a wave tries to blow into the low pressure centre but the Coriolis effect deflects the air to the right (in the northern hemisphere) to create an anti-clockwise circulation of winds. The effect of this wind pattern is to push the warm air which is to the east of the small low pressure centre more strongly against the cold air, so that the warm air is forced to rise even more than it would otherwise do simply by virtue of its proximity to the denser polar air. Similarly, the cold air to the west of the centre is pushed more strongly against the warm air and, being denser, undercuts the warm air increasing the tendency for ascent here also. The increase in ascent leads to an increased inflow effect, a stronger circulation and, as a consequence, more lifting of warm air over cold to the east of the wave and stronger lifting of the warm air by undercutting of cold air to the west. This, in turn, leads to increased inflow and so on, so that the wave, once formed, is a self-generating feature of the atmosphere. The ascent of air in the developing wave will be enhanced by the release of latent heat from condensation causing an increase in the buoyancy of the warm moisture laden subtropical air.

Occlusion

Fig. 8 shows a typical sequence of events as a wave develops into a major low pressure area and then decays as the cold polar air to the west of the centre sweeps right around to the south of the low and cuts off the supply of warm moist air to the low centre. This cutting off process is known as *occlusion* – a word which you will hear sitting in the dentist's chair when it has the same basic meaning. This prevention of any more warm air being available to push against the cold air, or to be undercut by cold air, means that the source of energy to the low pressure area is cut off. The low is then said to be occluded.

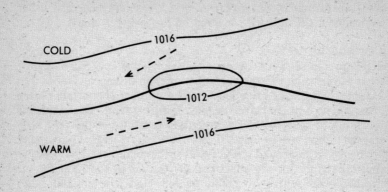

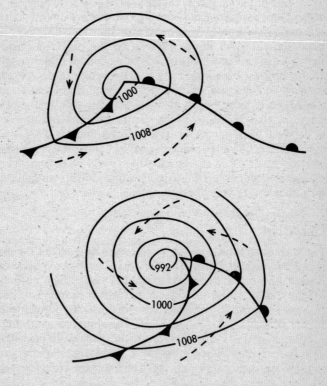

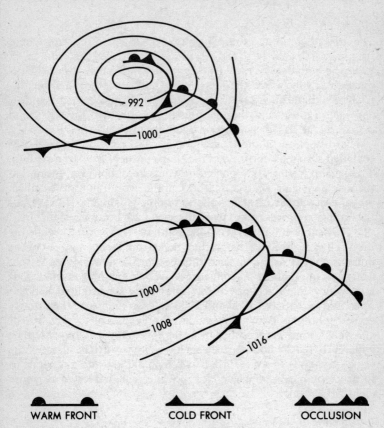

WARM FRONT   COLD FRONT   OCCLUSION

Fig 8 The development of a small ripple on the boundary between polar air to the north and air from the subtropics into an occluding low. The dashed arrows show the surface wind patterns. During this process the whole pattern moves in a west to east direction at a speed typically in the range 20–50 knots.

Warm and cold fronts

The area of cloud and rain which is formed where the warm air is rising over the cold air to the east of the low is called the *warm front*, being the boundary of the advancing warm air. Where the cold air is undercutting the warm, the area of cloud and rain is the *cold front*, this is where the cold air is advancing behind the warm. Where the cold air has come right round the low pressure centre then the two cloud masses merge together to become the occlusion.

The distortion of the frontal wave from a minor ripple to the occluded system is mirrored in the shape of the upper wind pattern. The general westerly component of winds does, however, mean that the whole system, while going through the evolution just described, moves eastwards with the oscillation in the jet stream. However, as the occlusion process takes place the jet stream becomes so distorted that it disrupts and re-forms to the south of the low centre while the low itself spins away to the north of the jet and, being now out of the influence of the strong upper winds, becomes a slow moving feature drifting around under the influence of its own spinning motion. This large whirlpool, which may by this time extend to a height of 30,000 ft or so and be between 1000 and 2000 miles in diameter, then slowly decays because of drag at ground level and eddies of various sizes at all levels.* It may be absorbed by a later low to form or may, itself, be apparently fed by another, newer, low. Either way the effect is the same – a merger of some form between an old low and a new one to form a complex low which, for a while, will have two or more identifiable centres.

Formation of an occluding low

The development of a minor ripple into an occluding low may take place within 24–36 hours and sometimes within 12–24 hours. This can be from the stage when the ripple is barely, if at all, discernible to the weather forecaster using conventional observations from land reporting stations and from ships on passage. The wave may, indeed, at this stage only be hinted at on pictures received from weather satellites. Many of the lows which affect the weather near the British Isles form initially to the north of the Caribbean or off the eastern

---

* Momentum of air is destroyed partly by means of drag caused by friction at ground level and partly by means of eddies – *eddy viscosity* is the term used to describe this effect.

seaboard of North America over areas of the ocean where data from conventional observational sources are frequently very limited. There is thus, often, an initial problem in the detection of the incipient waves and the almost impossible problem of deciding precisely where and when a wave will form before it reveals itself to the weather satellite cameras. Once a wave has been detected the problem then becomes one of forecasting not only its speed of movement but also its rate of development. Lows rarely maintain the same course and speed for long and, as we have seen, the occlusion process is associated with a slowing down of the low and a turning polewards (to the left in the northern hemisphere).

Errors, which are all too easy to incur, in the forecast of the occlusion process of a low will lead to errors in the forecast of the amount and rate at which the low will turn polewards. These errors can lead to quite sizeable errors in the accompanying forecasts of wind and weather. The effects may be especially marked in the vicinity of the British Isles because it is in this area that the turning of the maturing low often seems to take place. That this should be so is, partly, a result of the distance of the British Isles from the preferred source regions of these frontal lows and partly an effect of the British Isles being on the western edge of a large continent.

Cirrus cloud

Fig. 9 shows the wind patterns for a wave on the polar front at the surface of the earth and those at jet stream level – typically about 30,000–35,000 ft. Between the two levels there is a fairly smooth transition from the surface to the upper levels with the wind veering with a height ahead of the warm front and backing with height behind the cold front. We have seen that the warm subtropical air is sucked into the centre of the developing low until such time as the cold air sweeps right around to the south of the centre and cuts off the warm air supply. Warm air at low levels ahead of the cold front is pushed upwards by the undercutting cold air and, on being lifted, is then taken into the stronger upper level winds over the warm sector and lifted further in rising over the cold air ahead of the warm front. At the highest levels the air being lifted is subjected to the winds of the jet stream and so any cloud that is formed at these heights will be carried well away ahead of the low by these very strong winds. Air at jet stream heights is very cold and can hold only a small amount of water vapour and, therefore, when lifted, will only give rather patchy,

Fig. 9 Surface winds (dashed arrows) and high level winds (broad arrow) over a developing frontal low.

tenuous cloud known as *cirrus* (because of its hairlike, fibrous appearance, cirrus being Latin for hair). The first sign of an advancing warm front is usually the spread, from a north-westerly direction, of high level streaky cloud; Plate 3 shows a typical example of warm front cirrus. Aircraft flying at these heights often produce condensation trails from the vapour in the gases emitted from their engines. If the air is dry then the condensation trails will evaporate very quickly. If, on the other hand, the air is starting to rise, as it will be ahead of the warm front, then the cooling of the air will, eventually, cause the air to become saturated. In the interim period, while the air is approaching saturation, the condensation trails will evaporate very slowly, or not at all, and so the first sign of an advancing warm front, before the appearance of any cloud, may be the increasing tendency of condensation trails to persist or even to spread out, apparently creating a layer of cloud.

Cirro-stratus cloud

As the warm front approaches, so cloud at lower levels comes into sight being carried by winds less strong than in the jet stream and, therefore, not being blown so far away from the low as the high level cirrus. This lower cloud, being nearer to the main area of ascending motions of the low for a longer time than the cirrus, has had more time to be lifted. Also, having been at lower levels the air will have had a higher water vapour content initially. The result is much more uniform cloud than the cirrus, often in layers or sometimes 'solid' through a large part of the atmosphere depending upon the strength and uniformity of the ascending motions as well as the amount of moisture in the air. Thus the observer at sea level first sees the thin wispy cirrus which appears to become more of a uniform layer of, still, high cloud known as *cirro-stratus* and which may well have a halo around the sun or moon (see Plate 4). This layer of cloud thickens and lowers until the sun can just be seen as through ground glass (to use the usual textbook description) before finally disappearing. Further thickening and lowering of the cloud will be accompanied by rain. The rain may be slight and intermittent initially but will usually become heavier and more persistent as the front approaches.

During this time the wind will have been backing from a general westerly direction, before the cirrus hove into sight, to a southerly or, perhaps even, a south-easterly, increasing in speed the meanwhile. The barometer will have been falling, slowly at first and then faster as the front approaches. The passage of the warm front may be presaged by a heavier period of rain. As the front passes the barometer will stop falling, the wind will veer to a westerly again and the rain will ease off, turn to drizzle or, perhaps, cease altogether.

Warm sector

Behind the warm front we pass into air of subtropical origin but which has not been lifted by the action of either the cold or warm front although some slight lifting might have taken place simply by virtue of the proximity of the low pressure centre being generally an area of ascending air. The air of the *warm sector* – the name given to the region ahead of the cold front and behind the warm front – is thus very similar to the subtropical air discussed earlier in this chapter. Typically, it is cloudy with the cloud sometimes thick enough to give drizzle. If the sea is cold enough then it may well be misty with sea fog. Well away from the centre the cloud will break to give blue skies – all the cloud of the warm sector being in the lowest

2000–3000 ft. Over land this cloud often breaks, especially in the summer, to the lee of high ground. A wide, open, warm sector, that is one with the cold and warm fronts a long way apart, can result in some very fine warm days indeed to the east of Britain with the foehn effect over the Pennines or Grampians. Near windward coasts or hills, on the other hand, the forced lifting of warm sector air can give heavy drizzle and very low cloud.

In the warm sector the wind remains fairly steady in speed and direction, the latter being, typically, to the south of west and the barometer is usually steady or shows a slow fall.

The cold front

The cold front is usually a much more sudden transition than the warm front because the air being lifted by the undercutting effect of the cold air is now carried along the length of the front by the upper winds, see Fig. 9, p. 36. This is unlike the warm front where the high level winds carry the cloud ahead of the front. The depth of cloud capable of giving rain is, as a consequence, much more confined to the region of the front than in the case of the warm front.

As the cold front approaches there is, typically, a sudden fall in the barometer reading, the drizzle turns to rain, which may be quite heavy with thunder and the wind may back a little, say from west-south-west to south-west. The cloud thickens and lowers. After a short period of rain the clouds break, the wind veers to a westerly or north-westerly direction and the rain ceases abruptly. Because the jet stream is carrying the high cloud along the length of the front there is not an overhang of cirrus to any great extent behind the cold front, it is not a warm front in reverse. In the cold air, heating by the sea causes convection so that the cold front cloud is frequently a combination of the warm air being lifted by means of the undercutting effect and the onset of convection in the cold air. This is one of the reasons for the more sporadic nature of the cold front rain and the possibility of thunder. The most noticeable change in the transition from the warm sector to the cold air behind the cold front is the improvement in visibility associated with polar air masses. As the cold front moves away the cold polar air becomes showery.

How long the showery type of air will persist after the passage of the cold front will depend upon whether or not another wave develops quickly to the west. If a wave does not develop then the cold air can become sufficiently well established for the showers to become heavy and, possibly, thundery. If another wave does develop then the first indication to the yachtsman is that the showers start to become less intense and the convective (cumulus) cloud becomes less well developed vertically. That this is so is because, while the lows are areas of ascending air, there are areas of descent in the small mobile high pressure ridges between successive lows. The descending air, gentle though the motions are, inhibits convection from the surface. Forecasters talk about the convective cloud and the showers being 'damped down'. Plate 5 shows cumulus cloud becoming limited in vertical extent as the first sign of the next warm front appears with cirrus spreading from the north-west.

The normal sequence of events in a mobile westerly pattern of weather is that of warm front, warm sector, cold front followed by cold air and showers, then a brief spell of fine weather followed by the next series of fronts. If the low is already occluded then the warm sector will be missing and all that the sailor will observe will be warm front cloud and rain followed by the cold front sequence.

Blocking highs

The frontal sequence is sometimes interrupted by one of the mobile high pressure ridges developing to become, in effect, an extension of the subtropical high, the Azores high in the case of the British Isles, or of a part of the polar high, usually a part that has migrated to Scandinavia or Europe. The Azores high extension is more usual in the summer half of the year, while the Continental high extension is more of a winter phenomenon. These highs can cause even large low pressure systems to turn sharply northwards or, sometimes, southwards; they 'block' the usual west to east flow of the weather systems and are, not surprisingly, known as *blocking highs*. Such highs are often very stable, static or near static, features of the atmosphere and are responsible for the long periods of settled weather which occur from time to time. The November fogs and the spells of fine weather which we often seem to enjoy in May or September result from these blocking highs.

Cold front wave

Another variant on the normal theme is the development of a wave on the cold front. In this case, instead of the cold air showers before the next wave sequence, a wave forms very quickly on the cold front and develops, sometimes with surprising speed and vigour, to become a

Position	Wind	Weather/Cloud	Barometer	Visibility
A	West, force 2–3.	Small cumulus, perhaps first signs of cirrus spreading from the NW. Dry (Plate 5)	Has been rising, is now steady	Good.
A to B	Backing to SW. Increases to force 3–5.	Cloud increasing to become layer or layers of high cloud eventually covering whole sky. May be halo. Dry. (Plates 3 and 4)	Starting to fall, slowly at first.	Good.
B to C	Further backing to SSW or S, even to SSE. Increases to force 5–7, perhaps 8.	Cloud continues to thicken and lower. Rain starts, light at first, heavier and more continuous later. Patchy low cloud in the rain below the main cloud base.	Falling faster.	Deteriorating in rain.
C to D	May back a little further then veer to SW–W decreasing to force 4–6	Cloud lowers to become more or less uniform layer. Rain turns to drizzle.	Further fall then steadies.	Becomes poor, may be misty or foggy.
D to E	Stays fairly steady W–SW force 4–6	Well away from centre of low, some breaks in the cloud and dry. Near the centre, dull with drizzle.	Steady or a slow fall	Remains poor with mist or fog.
E to F	Backs a little and increases to force 5–7. Then veers to W–NW force 5–7, perhaps 8.	Cloud lowers, drizzle thickens and turns to rain. Heavy rain possible perhaps with thunder. Cloud breaks and rain ceases.	Increased fall then rise, perhaps quickly.	Poor then suddenly improves to good.
F to G	*Either* Veers a little more and remains force 5–7 or 8 with gusts. *Or* Remains in the West, backs WSW at times veers to WNW at others. Force 5–7 or 8 with gusts. *Or* Starts to back again.	Cumulus cloud develops with showers, possibly heavy and thundery. (Plate 2) A good deal of cumulus type cloud sometimes merging to complete cover. Showers, prolonged at times and, perhaps, heavy. Upper level layer cloud starts to increase again.	Continues to rise. Small rises and falls. Becomes steady then starts to fall.	Good except in showers. Good except in rain. Moderate or good.

The first sequence under F to G is when the next ridge of high pressure is advancing steadily. The showers will die out before too long and, after a short fine spell, the next front will approach. The second sequence is what happens when the next ridge is delayed and there are troughs of low pressure behind the cold front that are not associated with any frontal waves. This may change to the first sequence above or to the third which is the dangerous one of a wave forming on the cold front and developing quickly to become a vigorous low.

Position	Wind	Weather/Cloud	Barometer	Visibility
H	Variable or west force 0–3	Small cumulus, perhaps first signs of cirrus spreading from the NW. Dry.	Has been rising, is now steady.	Good.
H to I	Backs to south or south-east, some increase in speed.	Cloud increases to become layer or layers. Dry.	Starts to fall.	Good.
I to J	Continues to back to south-east or east. Increases to force 5–7	Cloud thickens and lowers to give rain, intermittent at first then continuous.	Falling	Deteriorating in rain.
J to K	Backs to north-east or north. Force 5–7.	Cloudy, rain at times.	Slower fall, Steadies then rises. No sharp change.	Moderate or poor in rain.
K to L	Backs to north-west or west. Then as in the three possibilities under F to G above.	As in sequence under the three possibilities of F to G above.		

Table 1 Weather sequences as observed from a yacht on the two tracks A to G and H to L, as shown on Fig. 10, when a low passes to the north or south respectively.

major low in its own right. Such lows can catch both the forecaster and the yachtsman unawares, but more of these unpleasant matters later.

The effect on the yachtsman

The sequence of weather described above was for a yacht to the south of the centre of a low; to the north of the centre the wind will back from the west through south then east to north-east and to north-west. Fig. 10 shows a low pressure centre and the associated fronts

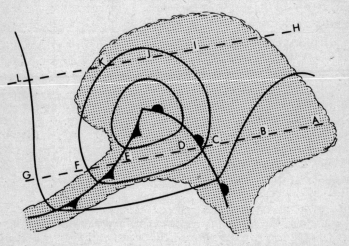

Fig. 10   Typical area of cloud associated with a frontal low. For weather along lines A to G (south of the centre) and H to L (north of the centre) refer to Table 1.

with two tracks of a yacht relative to the low. Table 1 summarises the various weather phenomena along the two tracks for ease of reference. The descriptions in this chapter and in Table 1 are very much of a 'by and large' nature. Fronts and other weather systems are like people – no two are exactly alike. The amount of wind change at a front, the extent of cloud ahead of a warm front, the duration and intensity of rain at a cold front can all vary enormously. At some fronts the wind can veer through almost 180 degrees, at others the veer is barely noticeable. This variability of the behaviour of weather at fronts and the uncertainty of the speed and direction of their movement may make life interesting for weather forecasters but, at the same time, they create problems for yachtsmen.

One last point to note regarding cold and warm fronts is that the terms 'warm' and 'cold' refer to the average temperature through a large depth of the atmosphere and not necessarily to the temperatures experienced at ground or sea level. Indeed, it may feel colder in a misty warm sector over the sea than in the clearer air behind the cold front. Over the land, where the cloud is more likely to break, the daytime temperatures in the warm sector are potentially higher than in the air ahead of the warm front or behind the cold.

Highs

Many textbooks on meteorology tend to concentrate on the weather associated with lows and mention highs only in passing as the quiet interlude between lows. This is understandable because lows are 'dynamic' features; they are the cause of the majority of the really bad stormy weather. They are newsworthy. The impression is often given that nothing much happens around highs and that they are uninteresting with no hazards. To some extent this is the case but variations in weather do occur of a rather subtle nature and can catch the unwary yachtsman by surprise. One of the reasons that it is not easy to describe the weather associated with highs is that much of the variation is topographical.

In predominantly maritime areas, such as the British Isles and its surrounding waters, the air circulating around a high has had a substantial period over the ocean. The air in the lowest 1000–3000 ft is, therefore, very moist. Also, the air in the high is descending and the cloud is confined to the lowest layers of the atmosphere because of the warming and drying out of the air during the descent. Low cloud will form wherever there is enough turbulent mixing to lift air to a level where condensation can occur and variations in the wind can cause variations in cloudiness within the circulation of the high. In itself this is of no great direct consequence to the yachtsman but the amount of cloud can obviously determine the amount by which the sun can heat the ground and, therefore, whether or not a sea breeze can develop. Even in a generally cloudy part of the high the passage of air over land, especially over hills, can cause breaks in the cloud by the foehn effect, as we saw earlier. The effect can be that on the windward coast a sea breeze will not develop because of insufficient insolation although a sea breeze might occur on a leeward coast.

The cloudy part of a high, particularly when the air is blowing on to a line of cliffs, may also be quite foggy or, at least, with some very low cloud.

When there is a sea breeze in the summer this can either result in an increase in the already existing onshore wind or, if the wind due to the pressure gradient is offshore, cause a reversal in the wind direction. What began as a force 3 onshore can increase to a force 5 while that nice spinnaker run out of Chichester harbour in a force 2 can become a beat to get to the south of the Isle of Wight. These sea breeze effects are more likely to occur in quiet anticyclonic (high pressure) weather when there are clear skies than in the more generally cloudy and, often windy, cyclonic (low pressure) weather.

A similar effect to the sea breeze, in a way, is the weather situation when pressure is low over land and high over the sea with the wind blowing along and, possibly, slightly on to the shore. If the sun can heat the land enough then the pressure over the land can decrease and the winds increase. This effect can be particularly marked with the heating over Spain during the summer when there is a seasonal low pressure area over the land mass. An extension north-eastwards of the Azores high gives north-easterly winds over the south of the Bay of Biscay. The central pressure of the low becomes lower by day and increases the winds over the Bay quite markedly; there is a corresponding decrease overnight when the pressure in the low pressure centre increases. The night time force 6 around Cape Finisterre frequently becomes a day time gale force 8!

3

Observing the Weather

In the first two chapters we saw how weather systems form and how they behave. The various types of weather associated with lows and highs and with different air masses were described. However, relating descriptions in a textbook to actuality is by no means straightforward and the yachtsman should make every effort to learn the skills acquired by the professional seafarer. In particular, he should learn how to recognise the basic cloud types and, above all, he should develop the habit of observing the weather and relating it to that which was forecast. Knowledge gained through experience is an everyday process; we soon learn to recognise that tapping under the bonnet of the car that indicates that attention is required to some vital part of the engine. Even when he is taking a spell down below the skipper of a yacht recognises that very slightly different feel of the movement of his craft that indicates a change of course or a slight change in the wind or sea. The level of experience needed for instant recognition of the warning signals of bad weather probably demands a higher degree of effort for the amateur than the examples quoted. Visual observations are of the utmost importance and greatly to be encouraged but it is, nevertheless, true to say that through lack of sufficient opportunity the amateur yachtsman is unlikely to approach the same level of ability as the professional seaman. The yachtsman is thus at greater risk than he may realise, a fact compounded by the relative frailty of his craft, by its lack of speed and, in some cases, by his own lack of experience in dealing with really bad sailing weather.

Yachtsmen, like all seagoers, are encouraged to make full and systematic use of the forecasts provided by the professional meteorological services of the world and techniques for recording and interpreting the shipping forecast are given in some detail in Chapter 6. However, even with the most sophisticated forecasting techniques available in the hands of competent forecasters the state of the science is such that forecasts do go wrong. Developments in the weather occur at different speeds from those expected or in a different manner. The sailor who is alert and watching carefully the behaviour of the cloud and wind may realise when there are significant differences from expectation and so will be warned that much earlier than the man who dismisses the whole business as too difficult or too complex. With these ideas in mind we will go on to see just what observations can be made reasonably easily and without taking up too much time from the main business of sailing.

Visual observations

The first type of observations are the visual, qualitative ones of the cloud and the weather. Is it raining? Are there showers? Is it drizzling? How much cloud is there? Of what kind? Is it increasing in amount or decreasing? Her Majesty's Stationery Office publishes the *Marine Observer's Handbook* which contains photographs of the main cloud types and there are many other books and charts which are worth studying; the Royal Meteorological Society, for example, sells some excellent wall charts of clouds. The meaning of the main cloud types and their relationship to the various weather sequences described in Chapter 2 should be clearly in your mind. The distinction between showers and rain or drizzle should be fully understood and, similarly, the significance of good visibility and poor visibility. If one of these phenomena is occurring then there must be a good reason which should be reasonably obvious from the forecast. The thinking sailor should continually ask himself 'Is what I see consistent with what I expect? And, if not, then why not?' The next step is assessing the implications of those observations not in accordance with the forecast. This will not be easy for most yachtsmen at first but constant practice of observing what is happening to the weather and comparing this with what the chart shows will pay dividends, and, before long, the various cloud patterns, associated with warm fronts, cold fronts, warm air and cold air masses, will become familiar.

Instrumental observations

The barometer

Leaving the qualitative, visual observations of cloud and weather we shall move on to those observations which can be made in a quantitative manner. The most important instrument, and one which is to be found on any yacht worthy of the name, is the barometer. The long glass tube filled with mercury to be found in school and other laboratories although an extremely accurate instrument is, clearly, not suitable for use on board a yacht. Rather more robust, but less sound from a technical point of view is the *aneroid barometer* which consists of a partially evacuated metal capsule surrounding a spring. Increases in the atmospheric pressure compress the capsule against the force of the spring while decreases in the pressure allow the spring to expand. These movements are translated by a system of mechanical linkages to the movement of a pointer around a dial. The linkages constitute a major weakness to this instrument and one which is compounded by the not so gentle tapping thought to be necessary by many people when reading the pressure. A spot of light oil on the linkages from time to time would probably help the performance of this really quite delicate instrument.

The aneroid barometer is not noted for its absolute accuracy, however, it is the *changes* in pressure in which we have most interest. Nevertheless, absolute values are of some use and it is worthwhile calibrating the instrument occasionally. The easiest way to achieve a reasonably accurate calibration is to wait for a quiet day in harbour when the wind is light and the pressure is changing only slowly. Read the barometer at about five minutes before the hour – any hour of the day will do but midday (GMT) is probably the optimum. About two hours or so later telephone a meteorological office and ask for the pressure at sea level for the place and time where and when you read the barometer. By means of a screw on the back you can then adjust your barometer by the difference of the two values. The choice of midday (GMT) is best because at that time the forecaster will have the maximum amount of data from which to draw his isobars and so estimate accurately the pressure at the relevant place and time. A day of slack winds and little pressure change also helps the forecaster make his estimate of pressure as accurately as possible. Be careful, incidentally, if you have read your barometer in, say, Chichester Harbour and then have travelled up to London before making your call to a meteorological office; the unwary forecaster may ascertain your whereabouts but may forget to ask – and you may forget to say – where the barometer is situated! If you take the barometer home before doing the calibration then you will have to tell the forecaster your height above sea level and ask for a pressure for that height; the barometer should then read correctly when you return to sea level.

Nowadays there is available a precision aneroid barometer which uses optical methods for the measurement of the expansion and contraction of the partially evacuated capsule thus removing most of the problems which exist in the conventional instrument. The precision instrument, which gives a digital read-out, is expensive but is capable of giving pressure readings comparable with those from mercury barometers while being more robust than the usual form of aneroid instrument.

Whichever instrument is used the barometer should be mounted near the chart table so that to read the barometer every hour is as automatic and as convenient as reading the log, which you will be doing anyway. All meteorological services of the world use the *millibar* as the standard unit of pressure and it is in your interest to have your instrument so calibrated. If it is not then make sure that you have a conversion card handy.

The barograph

The *barograph* is an instrument which, instead of having a needle moving around a circular dial, has a pen moving up and down the side of a rotating drum around which is wound a paper chart. This gives a continuous record of the pressure. In most small boats there is far too much movement for this instrument – which is more delicate than an ordinary aneroid barometer – to be of much use while on passage. A barograph is, however, useful in harbour when you are asleep or ashore. Some yachtsmen have found that if they mount a barograph on a thick cushion of rubber foam – athwartships and amidships is the best siting – then even on passage, a useable, if rather thick, trace can be obtained. The main benefit is still in harbour though.

Measuring air, sea and dew-point temperature

Temperature, sea temperature and dew-point temperature are normally not measured by yachtsmen and yet are easy to obtain and are of more than academic interest. The sea temperature is the easiest, being measured simply by heaving a bucket overboard on the end of a rope, bringing it back on board and quickly taking the water temperature before it warms up. Any reasonably accurate and robust thermometer will do although there are specially constructed and shielded thermometers designed for this purpose. Take care not to take a sample of water from near the engine outlet.

The air temperature and the dew-point can both be obtained from an *aspirated or whirling psychrometer*. These rather grand sounding instruments consist of two thermometers one of which has its bulb wrapped in muslin which is dipped into a reservoir of pure water (distilled, but free from battery acid). In the case of the whirling psychrometer the thermometers are mounted on a frame which is swung by a handle – rather like a football supporter's rattle. The aspirated psychrometer has the two thermometers shielded by a housing which also contains a motor driven fan. The object of both instruments is to pass a stream of air over the two thermometers so that the water on the muslin evaporates and, in so doing, takes heat from the thermometer bulb – the heat taken is the heat necessary for the water to evaporate. The drier the air the quicker the water evaporates, so that more heat is required from the thermometer bulb and the lower the temperature that the wet bulb will show. The wetter the air the slower the evaporation and the less heat that is needed from the thermometer bulb so that the two readings come closer together. If the air is saturated then no evaporation can take place and the two temperatures are identical. A set of tables supplied with the instrument enables the dew-point to be obtained from the dry bulb temperature and the depression of the wet bulb below the dry bulb reading. The main use of these readings to the yachtsman is to enable him to estimate the likelihood of sea fog; this will be discussed in the next chapter.

Anemometers and wind vanes

Wind can be included in the observations which may be made objectively using instruments which are found on many yachts today. Many yachtsmen prefer the subjective technique of estimating the wind force from the behaviour of the yacht and the state of the sea. First, however, the instrumental methods. *Anemometers* and *wind vanes* mounted at mast top height are finding increasing favour on large yachts these days with remote reading dials situated in the cockpit or near the navigation table. These are usually reasonably accurate although there must be some disturbance of the air by the mast and sails. The instruments should be sited on the tallest mast; readings from the mizzen of a ketch will be unreliable because of eddies from the mainsail and mainmast.

Ventimeters and compasses

Cheaper by far, not so accurate but good enough for many purposes, is the hand held *ventimeter* with a hand held *compass* and a *telltale* to give the wind direction. These should be held as far out to windward as possible and away from the rigging and superstructure. The ventimeter consists of a cone-shaped Perspex tube with a small orifice near one end and a light disk inside. The pressure created by the wind at the orifice makes the disk rise in proportion to the increase in the wind speed. At low wind speeds the ventimeter tends to read a little low but at about force 4 and above it is reasonably accurate. With both of these instrumental techniques it is necessary to make allowance for the speed and direction of the yacht because all that either can measure directly is the wind relative to the yacht. This calculation is probably best done graphically, in much the same way that the speed and direction of the yacht over the ground can be obtained from the direction and strength of the tidal stream and the speed and direction of the yacht through the water.

The Beaufort scale

In practical sailing terms more important than knowing the wind speed in knots is an appreciation of what effect the wind is having or is likely to have on your yacht and, for this reason, the traditional method of estimation is to be preferred. The technique was originally described by Admiral, then Commander, Beaufort in 1805 during the Napoleonic wars. The definitions, as set down by Beaufort, were remarkably precise, although this may seem strange to a non-sailor and, possibly, to a present day yachtsman as well. The descriptive terms 'light breeze', 'moderate breeze' and so on can, of course, mean all things to all men; a moderate breeze to a dinghy sailor on the Thames may be but a light zephyr to a yachtsman taking part in the Fastnet race. The precise behaviour of a man-of-war during the Napoleonic wars was well known to the officers and men of His Brittannic Majesty's Navy and the terms used by Beaufort were very clear. To have 'all sail set and clean full' meant to be sailing under full sail, a little off the wind so as to be making maximum speed, certainly not quite close hauled and most definitely not pinching! The phrase 'full and by' meant keeping the man-of-war as hard on the wind as possible, while being 'in chase' meant that the captain was cramming on all possible sail in order to make as much speed as possible and yet not run the risk of losing a spar. This was a matter for fine judgement, as are all races in strong winds. In Beaufort's day the prize was glory while failure could well mean the ignominy of a court-martial. As very many sailors could recognise the wind in these terms, and because the handling characteristics of ships of the line, on the one hand, and frigates on the other, were remarkably consistent from ship to ship, the scale of Commander Beaufort was an incentive to aggressive yet good seamanship.

The important point about the Beaufort scale is that it is the effect of the wind on the ship or yacht which defines the wind force and not what a person may feel or an instrument measure. In the man-of-war the man in the crosstrees would feel quite a different wind on his face than the officer of the watch on the quarter deck but each would be able to describe the wind in the same terms. Each would know whether or not it was safe to shake out another reef or hoist another sail. The scale had, incidentally, no need for decimal points or any finer discrimination. For forces 5 to 9, for example, the distinction between one force and the next was a matter of one reef more or less or one sail more or less. It is not possible to have half a reef or a sail half way up; a sail is up or not, a sail has no reefs, one reef, two reefs or more. There are no half measures. The highest point of the scale was a matter of relevancy; there is no point in being concerned about the precise wind force if, instead of wondering how to catch the French frigate, all that could be done was to run under bare poles and pray! Beaufort saw no need for further graduations either within his scale or beyond his definitions.

Modern dinghy classes can form their own version of the Beaufort scale. In a force 2, for example, the helmsman and crew of a Heron dinghy will still be sitting on opposite sides of the boat while in many other classes with higher performance the occupants of the boat will be on the same side. Hornets, International 14s and Merlin-Rockets, for example, will normally plane in a force 3 while it may be a force 4, or the very top end of a force 3, before a Firefly or a National 12 will plane. In a force 5 almost all racing dinghy helmsmen will be easing their sheets in the heavier gusts. The modern, deep

Wind force	Description	Wind speed (knots)	State of sea	Probable wave height (metres)
0	Calm	0–1	Sea like a mirror.	0
1	Light air	1–3	Ripples with the appearance of scales are formed, but without foam crests.	0
2	Light breeze	4–6	Small wavelets, still short but more pronounced. Crests have a glassy appearance and do not break.	0.1
3	Gentle breeze	7–10	Large wavelets. Crests begin to break.Foam of glassy appearance. Perhaps scattered white horses.	0.4
4	Moderate breeze	11–16	Small waves, becoming longer, fairly frequent white horses	1
5	Fresh breeze	17–21	Moderate waves, taking a more pronounced long form; many white horses are formed. Chance of some spray.	2
6	Strong breeze	22–27	Large waves begin to form, the white foam crests are more extensive everywhere. Probably some spray.	3
7	Near gale	28–33	Sea heaps up and white foam from breaking waves begins to be blown in streaks along the direction of the wind.	4
8	Gale	34–40	Moderately high waves of greater length; edges of crests begin to break into the spindrift. The foam is blown in well-marked streaks along the direction of the wind.	5.5
9 *	Strong gale	41–47	High waves. Dense streaks of foam along the direction of the wind. Crests of waves begin to topple, tumble and roll over. Spray may affect visibility.	7

Wind force	Description	Wind speed (knots)	State of sea	Probable wave height (metres)
10 **	Storm	48–55	Very high waves with long over-hanging crests. The resulting foam in great patches is blown in dense white streaks along the direction of the wind. On the whole the surface of the sea takes a white appearance. The 'tumbling' of the sea becomes heavy and shock-like. Visibility affected.	9
11 **	Violent storm	56–63	Exceptionally high waves (small and medium sized ships might be for a time lost to view behind the waves). The sea is completely covered with long white patches of foam lying along the direction of the wind. Everywhere the edges of the wave crests are blown into froth. Visibility affected.	11
12 **	Hurricane	Above 63	The air is filled with foam and spray. Sea completely white with driving spray, visibility very seriously affected.	14

Table 2   Modern sea-going version of the Beaufort scale.
*Note**   In the shipping forecast and gale warnings this is referred to as 'severe gale'.
*Note***   Forces 10, 11 and 12 are all referred to as 'storm' in shipping forecast and gale warnings.

sea version of the Beaufort scale relates the wind force to the behaviour of the sea; the state of sea is likely to be recognised by all seafarers regardless of the size of their craft – gone are the days when all seamen had a common standard, the crew members of a bulk carrier are not likely to have an appreciation of the handling characteristics of a modern day frigate in a force 9! The modern 'state of sea' version of the Beaufort scale, given in Table 2, is, in effect, a description of the integrated effect of the wind averaged out over the whole spectrum of wind speeds.

In addition there is also the landsman's version of the Beaufort scale in which the various criteria define wind force in terms of dust being blown, trees swaying, branches breaking and similar phenomena.

In a rather more light-hearted vein there are the two versions which owe their derivation to some original research by Michael Green and reported in his book *The Art of Coarse Sailing*. The sea-going version refers to such happenings as the blowing of tea towels off the rigging and the difficulties encountered when trying to make tea while under way. The landed based version uses such observations as the blowing of the froth off the beer and the difficulty met by elderly customers in opening the doors to public houses. Michael Green's version may not be of general applicability and should be used with care, the original work seems to owe a good deal to experience gained while sailing in the atypical conditions to be found on the Norfolk Broads.

All the versions of the Beaufort scale, Beaufort's original, the modern sea going, the dinghy, the landsman's and those due to Michael Green, describe the wind in terms of commonly observed, reasonably reproducible effects. A dinghy sailor knows when he is in marginal planing conditions, the observant landsman knows at what wind strength he can expect trees to start swaying and Michael Green knows, more or less, when the sign outside his favourite pub will begin to swing. The important point is to have an awareness of what effect a wind of a certain force will have on your boat and on the sea rather than to know how many knots it represents. Instead of using instruments to measure the wind it is better seamanship to develop an awareness of what effect the wind is having and then to use the instruments occasionally to help to calibrate you and your boat if necessary.

Other useful equipment

In this chapter on observations and instruments it is not out of place to include other equipment used by yachtsmen and which, although not strictly instruments, will be of assistance meteorologically.

First in order of priorities comes a *radio* capable of receiving the BBC long wave frequency 200 kHz (1500 m), the BBC medium wave frequencies, and the PO coastal radio stations (including those broadcasting on VHF).

A *tape recorder* is useful to help take down the forecasts and an *alarm clock* can remind you that a forecast is due. If you put your trust in modern electronic gadgets then you might like to invest in a radio/tape recorder/alarm clock combined as the ultimate luxury.

A *pad* to take the forecasts down is vital, that published jointly by the Royal Yachting Association and the Royal Meteorological Society is probably the most useful, together with writing and drawing equipment. When you come to try to draw your own weather chart from the forecast you will need an eraser, a soft pencil and dividers – but these should already be on your chart table.

4

Fog, Local Winds, Sea and Swell

Throughout the centuries seafarers have had to be knowledgeable in the ways of the weather and, indeed, not to be would court disaster. Today, even with the weather forecast services which are readily available by radio and other means, there is still a lot the yachtsman can do to apply his experience and intelligence putting important, and sometimes critical, detail on to the bare bones of the routine forecast and so improving upon the most accurate weather forecast which he is likely to receive. In this chapter we shall look at some of the ways in which this can be achieved.

Sea fog

Sea fog is probably the greatest hazard which faces the seagoer and yet it is a phenomenon which is not always dealt with in the forecast in sufficient detail to meet the needs of the sailor. In order to know as much as possible about the occurrence of fog the sailor ideally needs to know the sea temperature, air temperature and dew-point. When discussing air masses we saw that when the air is cooled, by whatever means, down to the dew-point temperature then cloud will form. If that cloud is at ground level we call it *fog*. Over land fog forms, typically, on a night with clear skies when long wave radiation of heat from the surface of the earth, at a time when there is no incoming radiation from the sun, causes the ground to cool. With absolutely no wind at all the cooling ground only cools the air in immediate contact with it and the resulting condensation forms as dew. If there is just a little wind then, because there is some mixing of the air in a shallow layer of air close to the ground, the cooling occurs throughout that depth of air and, although some of the condensation still gives dew, fog droplets also form. With rather more wind the fog forms a little way above the ground in the form of a uniform sheet of cloud known as *stratus*. In the case of cloud or fog the heat of the sun during the following day causes the water drops to evaporate and the cloud or fog disappears – or 'burns off' in the jargon of the forecaster. If the fog is dense and deep in vertical extent, as may happen after a winter's night with a long time for the cooling and condensation processes to take effect, then the sun may not have sufficient power or time to complete the burning off process. The next night, because the condensation process can start as soon as the ground and air begin to cool, the fog will be thicker and, so, on occasions when fog forms on several nights successively it will often be thicker each night than the preceding one. This *radiation fog*, as it is called, has a well marked diurnal cycle – clearing completely or partly by day and reforming or thickening by night.

In the case of sea fog the temperature of the underlying surface barely changes throughout the 24 hours, except in shallow water, and sea fog does not have much, if any, diurnal variation. The process at work is simply that the sea has the effect of cooling warmer air down to the sea temperature – it cannot cool the air any further. If the dew-point is below the temperature of the sea then sea fog cannot form because the air cannot be cooled sufficiently. Conversely, with air of polar origin, when the air is initially cold and dry, passage over the sea has the effect of warming the air to a temperature which may approach the sea temperature – it obviously cannot be warmed any higher. The dew-point will also increase, as water from the sea evaporates into the air, but the dew-point must be less than the air temperature or, possibly, more or less equal to the air temperature at most. In the case of air being warmed by the sea the dew-point will, in fact, tend to lag behind the air temperature. In an air mass of polar origin, therefore, the sea temperature will be above the dew-point and sea fog is a virtual impossibility.

When air comes from an area of warm sea towards colder water then sea fog becomes an increasing risk. Typically, this is the case when air of subtropical origin approaches south-west England. The air starts off over a warm sea and is itself both warm and moist, the actual dew-point value, which is a direct measure of the water vapour content of the air, being dependent upon the time which the air has spent over the warm water. As the air moves over colder water then the sea cools the air down towards, and often as far as, the dew-point. By measuring the sea temperature and the dew-point the sailor can estimate the likelihood of fog. If the two values are close then fog is possible.

There is, perhaps, not a great deal that the yachtsman can do in this situation except, perhaps, ensure that he has got some good, accurate, navigational fixes especially if he is anywhere near land. In some circumstances, however, it might be possible to set a course away from colder water and so minimise the risk of encountering fog. The *Metmaps* published by the Royal Yachting Association and the Royal Meteorological Society (see Fig. 11), and which we shall discuss later, have values of sea temperature printed on them that are average values for the early spring, when the sea is at its coldest, and for the autumn, when the sea is at its warmest. These values can be used to indicate just how the sea temperature varies from place to place around the British Isles at these two times of the year. For other times estimates of average values can be obtained by means of interpolation between the two seasonal extremes. The yachtsman might be able to use these maps to steer a course avoiding colder water or, possibly, to head for warmer water.

Sea fog, as has been noted above, often occurs in air coming from the south. Over the sea areas Biscay and Sole, for example, sea temperatures in the autumn are about 17–18°C while in sea areas Fastnet, Lundy and Irish Sea temperatures of 14–16°C are more likely. Fog is, therefore, quite a threat in these latter sea areas in a southerly airstream. On the other hand, around the western coast of Scotland in the summer, the sea temperatures are similar to those out over the sea to the west and south-west, that is in directions from which the air can approach the Western Isles with a sea track.

Fig. 11 Map of the Sea Areas around the British Isles. The dots with single letters show the locations of actual weather reports from observing stations included in the shipping forecast, these are tabulated at the bottom of Fig. 17. The pairs of numbers show the minimum and maximum sea temperatures at various places. The former occur in February and the latter in September.

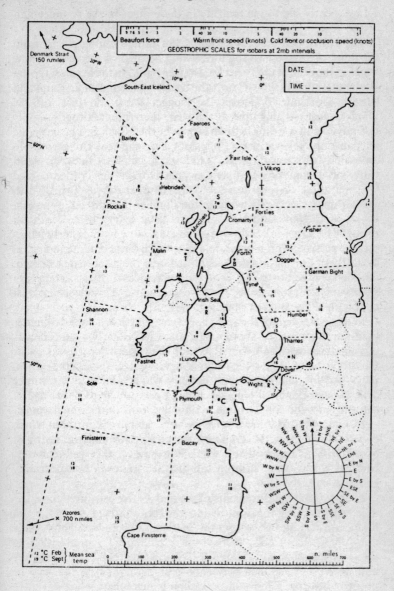

Consequently sea fog is unlikely in these areas in the summer half of the year.

The North Sea is an interesting and rather complex area from the point of view of sea fog forecasting because of the very shallow water in the central area near the Dogger Bank. In the autumn typical sea temperatures are 13°C near the coasts of north-east England and eastern Scotland, 15°C over the Dogger Bank and 16°C off the Danish coast. At this time of the year, therefore, sea fog is simply not possible in a westerly airstream over the North Sea because the air can only get warmer as it passes over the sea. Conversely, in an easterly airstream, the air is first made moist by the very warm water off Denmark and over sea area Dogger but is then cooled by some three or four degrees as it crosses sea areas Tyne, Forth and Cromarty. Easterly winds in the summer blowing on to the east coasts of England and Scotland thus have a very high likelihood of fog. The weather conditions often associated with easterly winds are the persistent high pressure areas which form over Scandinavia or northern Europe and this means that near the east coast of Britain, particularly the north-east coast, fog may occur for several days in succession and, sometimes, for a week or more. This persistent mist or fog on the east coast is called, variously, 'haar' or 'fret'.

In the winter half of the year the eastern North Sea becomes colder than the waters of sea area Dogger and so a westerly wind becomes one in which sea fog is possible over the eastern North Sea. Near the east coast of Britain the water is, again, colder than in the central parts and fog will occur with easterly winds which, in the winter also, are associated with persistent high pressure areas. Haar or fret can thus last for considerable periods in both summer and winter.

In winter there is some relatively warm water in the northern North Sea resulting from the Gulf Stream curling around the north of Scotland. This means that a wind blowing over this warmer water towards colder water inshore can give sea fog even in a northerly or north-easterly wind.

Even without measurements of sea and air temperatures considerable help for sea fog forecasting can be obtained from the sea temperature values shown on the *Metmaps*. These have some value not only for day to day tactical use in sailing but can also help in the strategic planning of cruises. If you have a choice about where and when to cruise then why choose an area which is likely to be prone to sea fog? Charts of prevailing winds together with average sea temperatures will enable you to determine in which areas sea fog is likely and, therefore, which are to be avoided.

Sea breezes

Emphasis has been placed on the dependency of the occurrence of sea fog on sea temperatures with their seasonal and geographical variability. The sea breeze is another phenomenon of interest to yachtsmen which also depends partly on sea temperatures. In this case it is the difference in temperature between land and sea which is the dominating factor because, like all winds, the sea breeze has its origins in the differential heating of the surface of the earth.

On a sunny day in the late spring or summer the land and, therefore, the air above it is heated more than the surface of the sea. The basic reason for this is that the sun heats only the first few centimetres of the ground which, if reasonably dry, can become quite hot. The sea, on the other hand, being a fluid and in perpetual motion, is able to mix the heat received through a large depth and so distribute the heat throughout a large mass of water which has, in any case, a higher specific heat capacity than the land. As a consequence, for the same amount of heat energy received from the sun, the sea will increase its surface temperature by only a very small amount compared to the land. Conversely, at night, when the surface of the earth is losing heat by radiation and receiving no incoming heat from the sun, the temperature of the land falls much more than that of the sea. Radiation of heat from the sea has little effect on its surface temperature, again because of the sea being perpetually in motion.

The consequential daytime warming of air over the land more than that over the sea results in an area over the land where the pressure is lower than over the nearby sea. This pressure difference gives a movement of air from the high to the low pressure. (In fact a completely scientific explanation of the sea breeze would have to say why the onshore winds at sea level were preceded by an offshore flow, at about 300 to 500 metres above the sea, from land to sea. The reason for this need not concern us here.) The overall effect is the development of a heat-driven circulation of air on a horizontal scale of a few miles with rising air over the land, a flow out to sea above the surface of the earth, descending air over the sea and a reverse flow from sea to land to complete the story. Initially, and in the absence of any wind due to a pressure gradient before the sea breeze starts to form, the wind of the sea breeze will start to blow directly onshore. As the sea breeze continues to blow the Coriolis effect starts to play its role, and the wind begins to veer. Fishermen talk about 'the wind going round with the sun'; this is simply an observation of the sea breeze being affected by Coriolis.

The strength of the sea breeze depends upon the driving force – the land to sea temperature difference which is, itself, dependent upon the temperature of the sea and the strength of the radiation from the sun reaching the ground. While the sea is still fairly cool in the late spring and early summer, but the land is getting warmer day by day, the sea breeze gains in strength. As the summer progresses and the autumnal equinox approaches, the incoming solar radiation starts to decrease but the sea temperature is now at its maximum. The temperature differences from land to sea become less marked and the sea breeze decreases in strength. By the winter half of the year, although the air temperature over the land may be high enough for a sea breeze effect to be noticeable from wind statistics, as far as the yachtsman is concerned it is a non-event. The sea breeze, for all practical purposes, is effectively confined to the period from April to September with May to August being the months of most likely occurrence.

On a sea breeze day the effect spreads inland as the land gets warmer and the sea breeze may reach 30–40 miles inland by the early evening in mid-summer, if the conditions are favourable. At the same time the sea breeze also extends out to sea by a distance of ten miles or so and up to twenty miles when the effect is at its strongest. The area of descending air out over the sea at the seaward extent of the sea breeze creates a mini-doldrums, on the landward side of which there might be a reasonably strong wind, especially near the coast. On a sea breeze day the yachtsman will do well to stand well inshore. That is usually where the strongest wind is to be found on a day of otherwise light winds; to go out to sea rarely pays because either you get into the very variable winds of the mini-doldrums or a rather light wind on their seaward side. In warmer parts of the world, such as the Red Sea, the sea breeze may extend considerably further out to sea and much further inland.

This description of the sea breeze assumes implicitly that the wind during the early part of the day is either very light or non-existent, and on many sea breeze days this will be so. When there is a wind arising from a pressure gradient then the effect due to the heating of the land acts as a modifying influence on the pressure gradient wind. For example, a south-west wind on to the south coast would back to a more southerly point under the effect of the sea breeze mechanism before veering back to south-west because of the Coriolis effect. An easterly wind would be veered by the sea breeze on the same coast to south-east and then further to south as Coriolis came into play. The effects on other coasts and other wind directions can be seen by adding graphically the two winds, the 'true' wind and the sea breeze wind, and then letting the resultant wind veer because of the Coriolis effect. The book *Wind Pilot* by Alan Watts describes the sea breeze in some considerable detail.

A strong wind caused by the pressure gradient, say a force 5 or more, will probably inhibit the sea breeze effect around British shores but with lighter winds and sufficient heating there will be some form of the sea breeze. The actual strength of the sea breeze does vary with the amount of heating but may reach a respectable force 4 around the coasts of Britain. The strong heating over the deserts of North Africa and the land surrounding the Red Sea is such that the sea breeze can knock a hole in a force 6 and can, itself, reach the same strength, or thereabouts.

The time of arrival of the sea breeze in the absence of any significant pressure gradient, is very regular and it has been claimed, for example, that the sea breeze reaches the Itchenor club house at opening time! The presence of a wind off shore will, however, delay the arrival of the sea breeze and in certain very critical conditions the pressure gradient wind and the sea breeze may reach a state of near balance so that the boundary between the two may waver to and fro for some considerable time. It is this kind of unreasonable behaviour which can be extremely trying for race officers in land-locked areas such as Poole Harbour.

The demise of the sea breeze takes place during the early evening as the land cools rapidly, which it does as the sun sinks in the sky and the ground receives much less heat than it loses. If the land cools sufficiently then the land and the sea can take on reversed roles and the air over the sea can become warmer than that over the land. If the temperature difference is large enough then there may be a land breeze blowing from land to sea for precisely the same reasons, in reverse, as the sea breeze. Around the coasts of Britain this land breeze effect is usually so light as to be barely discernible, but in other parts of the world where there is desert near the coast (desert cools rapidly at night), then the land breeze can have a very marked effect. The Red Sea is such an area and here the diurnal sequence of onshore winds by day and offshore by night is seen very clearly at the various meteorological reporting stations on the coast. With the prevailing northerly wind blowing down the Red Sea in the hot season any yacht wishing to make headway northwards can make use of this effect by taking a long offshore tack by night and a long onshore tack by day.

Katabatic and anabatic effects

Although the land breeze, as such, is barely, if at all noticeable around British coasts there can be a strong nocturnal offshore wind in some places because of the *katabatic effect*. This occurs when air is cooled by ground which is both cooling and slopes because then the cooling air can 'drain away' down the slope; if the slope is steep then this effect can be very marked. The effect is, of course, enhanced when the sea is warm. Strong katabatic winds can occur coming off the mountains around the Scottish lochs and the Norwegian fjords while there are particularly strong katabatic effects around the Adriatic.

The reverse effect to the katabatic is known as the *anabatic* which is the result of ground being heated which is sloping and facing the sun when areas nearby are not. This happens, typically, in mountain areas where one side of a valley may be in the sun while the other is in the shade. The result is a sea breeze type of mechanism with a flow of air at low level from the colder areas to the warmer ones and a reverse flow at a high level. Sailing on lakes in the mountains can be interesting or frustrating, depending upon your point of view, with these winds. As far as the sea sailor is concerned the anabatic effect is really only an enhancement of the sea breeze. A gentle slope facing towards the south might be expected to have a stronger sea breeze than when the ground is flat. On the other hand a very steep slope, such as a a line of cliffs, may act as a barrier to the sea breeze.

The sea and land breezes with their katabatic and anabatic aides are winds which depend for their existence on the differential heating of different parts of the surface of the earth.

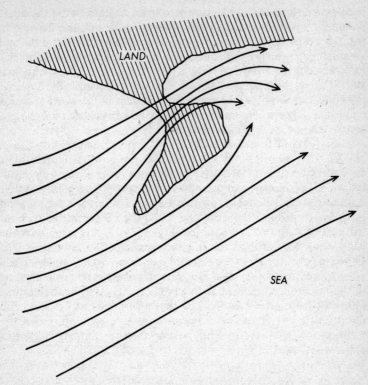

Fig. 12 Distortions in the air flow around Portland Bill.

Local winds

Other local winds of which the sailor will be well aware are the result of obstacles to the wind. The effects of headlands, cliffs, straits, and, on an appropriate scale, trees and houses can all cause variations in the wind speed and direction and the sailor should develop an awareness of this if only because they are unlikely to be mentioned in even the most detailed of weather forecasts. The variations in the wind arising from obstructions are such that no hard and fast rules can be given to help the sailor to know precisely what to expect. The general nature of the behaviour of the wind can be outlined so that he should be able to relate some of these peculiarities to their probable cause and so can be warned to know what to expect in similar weather patterns in the future. He might also be able to translate experience gained in one area to other similar areas and so anticipate possible problems when in unknown waters. With this knowledge if the wind is different from the forecast the yachtsman should have some idea as to whether this is because the forecast is wrong and there may be some development in the weather which will cause concern, or simply that the wind he is experiencing is the result of some local peculiarity.

The air, like all fluids, takes the easiest route around an obstruction and so, when blowing towards a headland or a hill, will often blow around rather than over such obstructions in its path. As a result the wind, apparently, can often be 'bent' by topographical features. A westerly wind blowing towards Portland Bill, for example, will be deflected around the headland while the air further out to sea will carry straight on, undeflected. Fig. 12 shows the result with the wind being bent around the Bill and 'squeezed' on passing it. The speeding up of the air in the squeeze is precisely what happens when air is forced to blow over a hill. The strongest winds over the top of the hill are the effect of such a squeeze, this time in the vertical rather than in the horizontal. To the eastern side of Portland Bill the opening out of the air trajectories gives a fanning out of the wind. If the wind is sufficiently strong then the fanning out effect

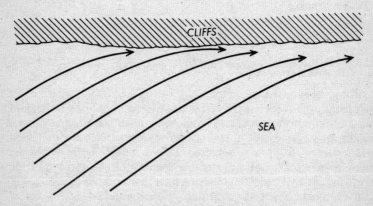

Fig. 13   The effects on the wind of a line of cliffs when the wind is blowing slantwise on to the shore.

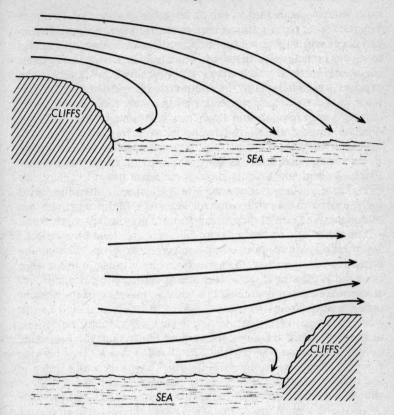

Fig. 14 The effects in the vertical when air blows across a line of cliffs offshore and onshore.

can form a complete swirl or a back eddy in a similar fashion to the eddies that you can see formed in a backwater of a river or a stream. Some headlands form particularly marked back eddies. One well-known one is that formed just to the west of Cape Finisterre when the wind is blowing strongly from the north-east on to the northern coast of the Iberian peninsula.

Even when there are no headlands the wind blowing slantwise on to a line of cliffs will be forced to blow along the coast in a direction more parallel to the coast than before and, at the same time, will be speeded up by the convergence of air on to the shore. Fig. 13 shows the air trajectories in such a case. A similar effect can occur when there are two headlands or coasts acting together to give a funnelling of the wind. The Strait of Dover is one cause of funnelling with which many yachtsmen will be familiar; this explains why winds from the south-west or north-east reach gale force in the sea area Dover but, perhaps, only force 6 or 7 in the adjacent sea areas up and down wind. A similar effect is often observed through the North Channel between Northern Ireland and Scotland when the wind is from the north-west or the south-east. Even through a channel as wide as St Georges Channel between Pembroke and south-east Ireland some funnelling effect is noticeable. The funnelling through the North Channel is often the reason for the issue of gale warnings for sea areas Malin and Irish Sea although over much of these areas the wind may be below gale force.

The bends and squeezes described so far are caused by the wind being deflected in the horizontal. Sometimes, however, the wind has no option but to go over an obstruction; one example of this is when the wind is blowing directly at a line of cliffs. With the wind from seaward then, on approaching the cliffs, the air is forced to rise some distance away from the coast. Near to the cliffs the wind can be quite light and there may even be a back eddy, this time because of the rotation of the air in the vertical rather than the horizontal as in the case of an eddy around a headland. Wind which is blowing from landward over a line of cliffs will, conversely, come down to sea level at some distance away to seaward of the coast. Fig. 14 shows some air trajectories in these two cases. In the offshore case there can be some quite strong winds where the splashdown occurs, stronger winds than those which occur further out to sea. An example

Plate 1 Cumulus cloud formed as bubbles of air rise from the surface of the earth. Watched carefully each cloud will be seen to be continually on the boil with bubbles emerging from the top, dissipating into the clear air and being replaced by new bubbles from below. Note the flat bottoms resulting from the level at which condensation occurs in the rising bubbles being constant over the area at this particular time (*C. J. Richards*)

Plate 2 The 'anvil' from a cumulo-nimbus cloud. This is caused by the spreading out of ice particles from the top of the cloud. (*C. J. Richards*)

Plate 1

Plate 2

Plate 3

Plate 4

Plate 5

of this will be known to many sailors familiar with Cowes and can be observed when the wind is blowing from the south-west over the hill behind the Royal Yacht Squadron. Dinghy sailors taking part in races from the Island Sailing Club start line will be aware of a dark patch of water with strong gusts around the north-east end of the line. The typical distance offshore of the splashdown effect is about ten times the height of the cliffs or hills.

Gust patterns

The topographical effects which we have been discussing are regular and, from experience, predictable features of the behaviour of the wind – not predictable in the sense that they should appear in shipping forecasts but predictable in the sense that the yachtsman should know enough about his home waters to know what to expect in the way of sea breezes, bends, squeezes, eddies and the like. By looking at weather maps and navigational charts he should also have some idea what to look for in strange waters. There are other localised variations in the wind which are of a more random nature and some of these are the completely random fluctuations in the wind caused by turbulence generated by the wind blowing over a rough terrain, the bane of the inland dinghy sailor. Little rhyme or reason can be attached to the gust patterns resulting. Nevertheless, there are some gust patterns which can be recognised and the

Plate 3   Cirrus streaks of the form which might be seen ahead of a warm front. These are composed of ice particles forming at the right hand end of each streak as viewed in this picture. If the wind did not change with height at these levels then the streaks would be vertical. However, as they fall the ice particles are carried by winds less strong than at the level above. The streaks do, in fact, show the change in the wind through the layer. (*S. D. Burt*)

Plate 4   A halo seen in cirro-stratus. If a warm front is not moving too quickly then the cirrus cloud shown in plate 3 can form a very uniform layer which gives rise to a halo around the sun or moon. These are the 'large' haloes referred to in Chapter 7. (*S. D. Burt*)

Plate 5   Here we see cumulus cloud becoming ragged and starting to flatten out as cirrus spreads across the sky ahead of the next warm front. (*C. J. Richards*)

behaviour of which can, to some extent, be predicted. These are the gusts which occur in convective weather conditions, for example those associated with cold air masses.

Convection takes place in the form of large bubbles rising from the surface of the earth because of local heating or because of a slightly warmer patch of water. As the bubbles rise away from the surface there are compensating areas of descending air. The height to which the rising bubbles ascend depends upon the amount of heating at ground or sea level and the vertical temperature structure of the air mass. The height from which the descending air originates depends upon how high the rising bubbles go, the more vigorous the convection the higher go the rising bubbles and the greater the height from which the descending air reaches the ground.

To understand how convection produces gusts we must return to the relationship between the wind and the pressure gradient. The wind blows more or less along the direction of the isobars and with a speed which is dependent upon the spacing apart of the isobars. Near the surface of the earth friction reduces the speed of the air and this slowing down affects the balance between the force on the air due to the pressure gradient and the deflection effect of Coriolis. Not only does friction slow the air down but it also causes the air to blow slightly across the isobars from high to low pressure. The amount of cross isobar flow is greater when the ground is rough and there is a marked slowing down of the air while over the sea, which is fairly smooth, the deflection is far less – 30 degrees or so over land and 10 degrees over the sea are fairly typical values. Above the surface of the earth the frictional effect decreases so that by about 2000 ft over the sea, or about 3000 ft over the land, the wind is blowing at the full strength implied by the spacing of the isobars and with a direction which is virtually along them. The wind at these heights is, therefore, both stronger than the wind at ground or sea level and is veered relative to the surface wind (in the northern hemisphere; it would be backed in the southern).

Returning to our convective weather situation we can now see what causes gusts. Air which has been at or near the surface of the earth for some while will be friction influenced and will move slightly across the isobars over the sea and rather more definitely across them over land because of the greater friction. If that air leaves the surface in a convective bubble it will be replaced by air which has come from some way up in the atmosphere and which will, consequently, be moving not only faster but will also be veered relative to the original surface air. The descending air retains its speed and direction during the descent (as long as the time taken for the descent is not too long). In time the 'new' air is, itself, slowed down by surface friction and so is forced to move across the isobars, in other words it becomes backed relative to the direction it had when it first reached the surface. That air in turn will be heated and forced to rise, only to be replaced by air moving faster and from a veered direction. The overall result is an alternating sequence of gusts and lulls with wind directions in the gusts being veered on that in the lulls. In very vigorous convection the descending air will originate from greater heights than the 2000 or 3000 ft mentioned above and the gusts will be correspondingly stronger. On occasion gusts can be double the average wind speed.

The time between successive gusts and lulls is, typically, a few minutes or so and a large yacht will barely be able to react although tacking on the right gusts and lulls is very much a skill developed to a fine art by successful dinghy helmsmen – tacking on to starboard on the gust will give a 'lift' up to windward and so will tacking on to port on the lulls. In the case of the large yacht in a day of strong gusts it might be prudent, if on the port tack, to sail just a little bit free so that gusts with their veer will not result in the jib being caught aback. On the starboard tack, on the other hand, the helmsman will be able to keep the boat pointing high up to wind secure in the knowledge that any strong gusts will veer and give him short-lived lifts up to windward.

Over the sea the barely changing sea temperature from day to night has the effect that the convection continues with much the same strength by night as by day. This is in contrast to the land where, because of the cooling of the ground at night, the convective activity quickly dies away as the sun goes down. The result is that not only do the gusts decrease very markedly at night over land but the overall wind strength also decreases. The lack of interchange between the air at the surface and that at higher levels in the atmosphere means that the air near the ground is entirely friction dominated. Even when the pressure gradient would imply a wind of 10 to 15 knots or so the effect of night time cooling can create a calm, especially if the sky is clear of cloud and there is rapid cooling of the ground accordingly. The process is aided if there are hollows where the air being cold, and therefore dense, becomes so sluggish that it cannot easily be moved even when there are strong winds above the friction layer.

. Gusts in convective weather that do not obey the usual rules are to be found near heavy showers. In such cases the falling rain or hail drags down to the surface of the earth the air from the level from which the rain or hail is falling. The weight of the precipitation is such that it falls quickly and so drags air, also quickly, down to the ground where it spreads out in all directions from the cloud. Before the rain falls the strong updraughts in the cloud will have been the cause of an inflow of air towards the cloud near the ground and, so, with these very heavy showers the observant sailor may notice a light wind blowing towards the cloud even when the cloud is moving towards him and then, just before the rain, a quite strong wind blowing away from the cloud with some very strong gusts. Depending upon just how quickly the cloud is moving the first gust may occur some five minutes or so before the rain arrives. The light wind blowing towards the cloud gives rise to the saying about the very heavy showers 'coming up against the wind'. They are, but the wind is, in effect, being created by the shower cloud itself. The strength of the gusts with these showers is a combination of the spreading out of the rapidly descending air and the fact that the air has come from a height where the winds are almost certainly stronger than the winds at sea level.

Near land, in the estuaries of rivers for example, the rough surface of the ground will make the wind very variable in a random manner with gusts which have no obvious pattern of behaviour except in so far as certain obstructions can clearly be seen to create turbulent eddies downwind of them. The sailor must be just that little bit more careful when sailing downwind of large oil storage tanks and suchlike when the wind is strong.

In warm sector air masses there is little or no convection over the sea and none over the land unless the ground can get hot enough. There is thus little interchange between the air at sea level and that higher up in much the same way as when the convection dies away over land at night. The result, here also, is that the air at sea level is entirely friction dominated but, of course, because the sea is fairly smooth this does not stop the air moving completely, as can happen over land on a cold night. The wind in the warm sectors of lows or in warm air masses generally is, thus, fairly steady and, for a given pressure gradient, lighter than that in a cold air mass. A pressure gradient which might give a force 7 or 8 in a cold air mass may just about reach a force 6 in the warm sector. The only vertical mixing of the air is due to turbulence, and unless the sea is very rough, this will be slight. If the sea is rough then the wind in the warm sector will have more gusts than normal but the strength of wind in this case would be such that to talk about stronger or more marked gusts would be academic.

The effects of the different air masses and surface roughness on the wind speed to pressure gradient relationship and the effects of the vertical stability of the air are such that there is not a precise one to one equivalence between pressure gradient and wind. The isobar spacing and direction are good indicators of the wind in a 'by and large' sense. The effect of gusts on both wind speed and direction means that, even over the open sea, there is no such thing as 'the wind'. There is a range of wind speeds and directions with the latter varying some 30 degrees or so in convective conditions and wind speeds of force 3 and above. Add to the convective effects the varying effects of topography, sea breezes and the like, as well as minor variations in the pressure pattern, then it becomes self evident that even to describe the wind actually occurring over a large area, such as the Channel, becomes difficult. The problems of trying to assess what will happen before the event might seem to be almost impossible. We will return to this when thinking about the shipping forecast in which the forecaster has to describe in a meaningful manner the expected behaviour of the wind over large and often inconveniently shaped areas for a twenty-four-hour period.

The standing wave

The standing wave is a wind phenomenon much used by glider pilots and its inclusion may seem to be out of place in a book on weather for sailors, but there are some instances when the standing wave can make life awkward if not unpleasantly dangerous for the yachtsman. The standing wave occurs when the wind is blowing across a range of hills which are lying more or less at right angles to the wind direction. There are certain conditions regarding the variations of wind and temperature with height that are necessary for the formation of the standing wave but we need not discuss them here except to note that they normally occur ahead of a slow-moving warm front or in the warm sector of a low.

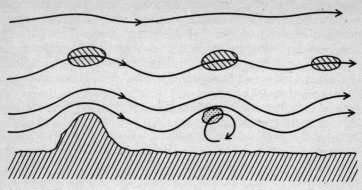

Fig. 15 Standing waves as the wind crosses a line of hills. Clouds form in the crests of the waves; those at low levels (1000–3000 ft) are often turbulent in appearance while those higher up (anything from 10,000 to 40,000 ft) take a smooth ovoid or lenticular form.

In the standing wave the air on crossing hills or mountains forms a series of waves over and downwind of the hills. These waves remain stationary relative to the hills with the air moving through them. In moving through the waves the air is forced to rise up to the crests and then descend into troughs successively and, if there is sufficient cooling as the air ascends, clouds are formed in the crests of the waves. As the air descends on the downstream side of the waves the air warms and the cloud evaporates so that the process is one of continuous formation on the upwind side and evaporation on the downwind side of each cloud. The clouds themselves are a very smooth ovoid form rather like a lens and hence their description in cloud atlases as *lenticularis*. There are often several such clouds in a train downwind of the hill responsible for the standing wave with each cloud in the crest of a wave.

In standing waves the atmosphere will be moving up and down throughout a large part of its depth under these orographic clouds. Where the air is rising the effect at ground or sea level is an area of light wind; in the extreme case there may even be a complete reversal of wind as a rotor forms. Under the troughs the wind may be quite strong because here the air trajectories are being squeezed together by the ground – see Fig. 15. It was this kind of effect which gave rise to the Sheffield gales in 1962. This form of wind variability can be found near the Isle of Man when the wind is from the east or the west. There is little that the yachtsman can do except that if he sees the wave clouds when he is some little way downwind of a range of hills then he should be on his guard for a local increase in the wind; if he experiences an area of light and rather variable winds then he might also find some areas with a wind stronger than the average on that day. The effects can be felt up to twenty or more miles downwind of the hills responsible.

Some of the topographic effects mentioned in this chapter can be very marked in areas of the world where large temperature differences occur between land and sea. The local winds of the Mediterranean, such as the Mistral, have their milder and relatively innocuous counterparts around our shores. Both the Bora of the Adriatic and the Mistral are essentially downslope funnelling effects and both can reach gale force.

Waves

Wind is important to the sailor because it is his prime source of energy and, if it becomes too strong, a prime source of danger as well. The wind also gives energy to the sea and creates waves which, in their own right, can become another source of danger to all sea-farers. How and why do waves form and what are the reasons for some of the really rough seas which occur around our coasts?

In the first instance wind blowing across water is subject to drag due to friction, even when the water is smooth, and the drag creates very small waves or ripples at right angles to the direction of the wind. Being raised just above the level of the water these small ripples present a surface to the wind. The force of the wind on the ripples then drives them forward simply by virtue of the pressure of the wind on their upwind side. However, as the waves cannot move at the same speed as the wind, and in fact move considerably slower, the wind continues to exert a pressure. This continued pressure causes the waves to grow as they run forward (grow, that is, in height and length with some increase in speed). The size to which the waves can grow depends upon the wind speed, the fetch of the wind over the water, the time for which the wind has been blowing and the depth of water.

At first the waves will be low and steep, that is to say that, although of small height, the ratio of height to length will be large. At some later stage the waves will have grown and become both higher and longer but the height to length ratio will be smaller than before. For any given wavelength there is a maximum speed for the wave and so, when a wave has grown to be as large as possible for a given wind speed then it will still be travelling at a speed less than the wind. The wave then breaks because the wind is, in effect, trying to push the wave along at a speed faster than its size and wavelength will allow. After breaking the wave will rapidly decay.

Waves appear to travel in groups. The upwind members of the group are those waves which are still growing, those in the middle have reached their maximum height and are just beginning to break while those downwind, or leading the group, have broken, are decaying and are thus starting to travel more slowly so being overtaken by their immediate upwind successors. The wind as we have seen is never steady and so waves are continually being formed by winds of differing speeds. This, coupled with the different speed of individual waves at any time of their short lives, means that at any instant we have a spectrum of waves of different heights and wavelengths all travelling at slightly different speeds. Because there are large numbers of waves with their individual speeds then precisely what happens at a specific location at a particular time depends upon just what waves arrive at that time. If a number of wave crests arrive simultaneously then they reinforce each other, as it were, and a large wave can result. If crests and troughs arrive at the same time then they can cancel each other out and, even in the roughest sea, there can be a brief period of relative calm.

Because of the dependence on random chance for the formation of the very large waves and because the individual components have their own speed of travel it becomes reasonably obvious that the giant waves experienced in the worst storms can have only a brief moment of glory. The probability of having waves of various heights can be calculated for a sea whose characteristics are known. It can be shown, for example, that one wave in twenty-three is twice the average height, one in 1175 is three times the average and one in 300,000 is four times the average height. Because wave components seem to take some time to get into phase to form these giants of the seas there often seem to be several waves of increasing size before the really big one, which is then followed by a number of decreasing size.

Out in the open Atlantic waves can be quite steep, but if they approach shallower water or an opposing current or tide then they can steepen dramatically and begin to break. In the case of waves approaching shallower water the reason is that the wave motion extends downwards in the ocean to a depth approximately equal to the wavelength of the waves. As the waves approach water of a depth much less than the wavelength then the waves become, in effect, too high for the depth of water and the waves begin to break. The breaking is caused by the excess energy in the waves; the depth of water determines the maximum wavelength and, therefore, the maximum kinetic energy which can be contained in waves in that area. If waves approach with more than the maximum energy for the depth of water then that energy of movement is converted into making the waves higher so that they become too steep and break.

The same effect occurs when waves are opposed by a current which causes a build up of wave energy which can only be realised, again, by an increase in the height of the waves to a degree such that the height is too great relative to the length. A current which is about a quarter of the speed of the waves will stop the progress of the waves completely and no energy will be propagated any further in the direction of movement of the waves. This is the reason for the rapidly steepening and then breaking seas when the tide turns in some of the tidal races around our coasts.

The converse of each of these effects also occurs when waves run on into deepening water or the tide turns and runs in the same direction as the wave movement. From being in a very rough and turbulent sea the transition is sudden and dramatic as the wave energy is absorbed by the deeper water or taken away by the current more rapidly than the waves are moving.

Sailors often talk about the effects of wind over sea and wind against the sea. Strictly speaking what is really meant is waves with or against the tide or current although, because waves are wind-generated, it is often the same thing in practice. Just occasionally a sea can still be running although the wind direction may have changed and if the waves are contrary to the tide then the rough sea will continue.

Yachtsmen sometimes ask why the weather forecast cannot give the state of sea. It should now be clear that this is virtually impossible to do in any meaningful fashion because of the dependence upon tide and water depth. At best all that could be given would be wave heights for the open sea and these can be read from the Beaufort scale state of sea description (Table 2, p. 52). These would have very little relevance when applied to coastal waters. In the same way that the yachtsman should develop an awareness of the various possibilities in the way the wind can vary he should also become aware of the effects of waves moving into shallow water or when waves are subjected to a change in the direction of the tidal stream. Experience, talking to local sailors and other seamen, and the use of common sense in translating experience from one area to another, are all ways in which the yachtsman can avoid being surprised by rough seas.

Swell

So far we have been talking about waves generated by the wind in the area where they are being formed – these are sea waves. When the waves cease to receive energy because the wind has died away or because the waves have moved out of the area where they were being formed they carry on travelling under their own momentum. Having lost their energy source the waves decay slowly with their height decreasing and their wavelength increasing. These wave trains are encountered in areas well away from where the waves were first formed and are known as *swell*. Swell can travel many thousands of miles before becoming imperceptible. The swell from storms off the coast of South America can be found off Nigeria. Swell is of a different appearance to wind waves in that it has a smooth and unbroken oily look while wind waves always break. The existence of swell can presage a new wind system, especially if the swell is coming from a different direction from the wind waves currently being experienced.

Wave size

How big can waves be? The modern version of the Beaufort scale describes the sea in terms of the wind force and Table 2 (p. 52) gives some typical heights in the open sea. Lawrence Draper in an article in the magazine *Motor Boat and Yachting* in 1971 quotes 70 feet as a fairly normal maximum wave height in the North Atlantic and the northern North Sea, 50 feet off Lands End, 30 feet in Morecambe Bay, the southern and central North Sea and 60 feet to the east of Peterhead. These values show the effect of shallow water in limiting the height of the maximum waves. The highest reliable observation of a wave is about 110 feet in the North Pacific. The incorrectly named tidal waves known as Tsunamis are generated by earthquakes and have been known to reach over 200 feet in height.

5

Shipping Forecasts and Gale Warnings

Weather forecasts are almost always expressions of probability regarding the general type of weather likely to occur over areas which may be large in some instances and quite small in others. For example, the national forecasts on Radio 4 for the United Kingdom describe the weather over such areas as the Midlands or Scotland; on the other hand some of the forecasts on VHF radio or on the PO telephone weather service deal with much smaller areas, sometimes just a city. Whichever kind of forecast you hear on the radio or read in the press there will be phrases such as 'rain may reach . . .' or 'showers, perhaps with longer periods of rain'. The use of such phrases does not mean that the forecaster is trying to avoid making a decision or that he is not trying to be helpful. The reason for such apparent prevarication is simply the sheer impossibility of being more precise for areas such as those mentioned, even the smaller ones. Indeed, on many occasions, to be more specific in the writing of a forecast would be misleading. The weather is just not sufficiently uniform over large areas nor does it ever behave in precisely the manner expected as regards speed of movement or development of the various types of weather system, that is the fronts, the lows and the highs.

In the USA weather forecasts for the general public cover the uncertainties by means of probability percentage expressions, for example '30% probability of rain'. The same technique is also used the world over in the specialised forecasts for landing conditions at airfields both civil and military.

General forecasts

In this country the current practice in forecasts for the public is to make use of the subtleties and nuances of the English language to convey the various possibilities of the particular weather situation. Even in the brief time allotted to the forecasters appearing on the television the person presenting the forecast can usually give some idea of the likely weather developments and indicate the uncertainties – will the cloud break early enough to give fog or not, will the thunderstorms over northern France spread to the South Coast or not? The general forecasts broadcast on radio and television are flexible in form so that the forecaster can vary the order in which he deals with the various constituent parts of the United Kingdom depending upon the weather situation and its complexities. If the weather over Scotland is likely to be more difficult to describe in the forecast than elsewhere then he may decide to give that part of Britain more time than the rest put together. He might be able to cope with remaining areas in a few throwaway lines.

When on land and having an involvement in meteorologically sensitive activities it is possible to supplement radio or television weather forecasts by means of personal telephone calls to a meteorological office, or even a visit to a weather centre established for just this specific purpose. At sea, unless you are on board a yacht equipped with ship-to-shore radio when it is possible to speak via PO coastal radio stations or direct, by R/T link, to the shipping forecaster at the Central Forecast Office at Bracknell, you have to rely upon the various broadcast services that are available. These services include those broadcast by the BBC on its general services, the PO coastal radio stations, local radio services from other countries and, if you can read morse, some specialist broadcasts for merchant shipping and the Royal Navy. The most useful of all the forecasts readily available to you is the shipping forecast broadcast on 200 kHz – 1500 metres, long wave. This forecast will provide the framework within which all the other forecasts which you may hear from time to time can be used and it will set the scene against which the thinking yachtsman should seek to interpret and understand the weather which he is experiencing.

The shipping forecast

The shipping forecast differs from almost all the other forecasts issued on the radio or television in that the form of the forecast is defined extremely precisely. Every sea area has to be mentioned in a predetermined order; many of the words used are given strictly defined meanings – virtually a code, in fact. There is little or no time to express doubts or other possibilities and the forecaster has no opportunity to do other than make up his mind as to what is the most likely development of the weather patterns of isobars, fronts, lows and highs and then describe the accompanying weather. The forecaster cannot indicate the possibility of alternative developments, all he can do is to express some doubts concerning the forecast wind and weather by means of such words as 'may' or 'perhaps', but even here the limitations of time restrict the possibilities open to him. Add to the uncertainty of the overall weather pattern the spatial and temporal variability discussed in the last chapter and it becomes readily apparent that the sailor, in order not only to be able to understand the forecast but also to be in a position to anticipate when the forecast is going to be seriously in error, has to have a good understanding of the weather which he should be experiencing and what variations he can expect because of the topography, sea breezes and so on. He should be able to form an opinion as to when the forecast is apparently in error or when the differences between what he sees and what he has been led to expect to see are not significant.

To achieve a degree of competence and confidence in assessing the weather forecast along these lines is not easy and demands a certain amount of skill combined with a good deal of practice. Just how much the average yachtsman can achieve is a matter for conjecture. The skills needed are those of being reasonably observant coupled with the ability to comprehend some simple physical principles. The time required for acquiring this facility is probably rather more than that needed to become a competent navigator. The practical difference between the two is, probably, that to be able to use the meteorological services in an adequate fashion demands reasonably regular and frequent practice while navigation, once learnt, is perhaps less easily forgotten. By way of compensation it is undoubtedly easier to practise meteorological skills in the warmth and comfort of your own home than it is to practise navigational techniques with any degree of realism.

Gale warnings

Before discussing the shipping forecast itself it is first necessary to talk about the gale warning service if only because the forecaster has to include within the body of the forecast a summary of any areas with gale warnings in force.

Gale warnings are issued whenever it is expected that the wind will reach at least force 8 anywhere within a given sea area during the next twelve hours. Even if, in the shipping forecast, the forecaster uses words such as 'locally' or 'perhaps', or some other word which implies a measure of doubt as to the extent or even the possibility of the gale actually occurring, then the warning still has to be issued. The warning is intended to draw the attention of the mariner to the possibility of strong winds, without necessarily implying that any individual craft will be affected, or even that the gales will definitely occur. The warning is, in effect, just another forecast and, like all forecasts, is an imprecise statement about the future.

If the strong winds are expected within six hours of the issue of the warning then the gales are said to be 'imminent'. Similarly, any change in the strength or direction of a gale expected within this six hour period is said to be imminent. Gale warnings always state the time at which they were issued from the Meteorological Office because there is inevitably some delay between the issue of the warning and its broadcast by the BBC or the PO coastal radio station; *it is from the time of issue that the imminent period refers and not the broadcast.* Gale warnings expected to come into force between six and twelve hours from the time of issue are qualified by the word 'soon'.

Sometimes a forecaster will wish to issue gale warnings for the period from 12–24 hours from the time of the gale warning and, in this case, the code word is 'later'. However, difficulties in forecasting are such that later gales have a considerable uncertainty factor and are only issued when the forecaster has a reasonably high degree of confidence that they will, in fact, occur. Forecasters would like to be able to issue all warnings of gales as 'later' and so give the maximum amount of warning but this is just not technically possible at the present time and forecasters have an understandable wish not to cry wolf too often. Therefore, there is a tendency not to issue warnings for the later period but to make use of a convention which allows the forecaster to use the words 'perhaps gale later' in the shipping forecast and, yet, not issue a gale warning. By this means the forecaster can indicate that he is considering a gale warning but has not yet got sufficient evidence to issue one. On hearing this phrase the sailor can be advised to be on the lookout without the forecaster issuing a warning which may cause premature action and which might not subsequently be justified. It should be emphasised that the 'perhaps later' facility does not absolve the forecaster from the responsibility of issuing a warning for the 'later' period if he has reasonable grounds for thinking that the gale will materialise. Neither, as stated earlier, does uncertainty in the first twelve hour period preclude the necessity to issue a full gale warning. The forecaster *must* issue a gale warning for this period, even if he qualifies the shipping forecast with considerable doubt by means of a phrase such as 'perhaps locally'.

An expected change in an existing or expected gale can also be the subject of a warning. If the wind strength is expected to change by one or more Beaufort forces and, at the same time, remain at or above gale force 8 then this can be expressed in some such words as 'gale 8 expected to increase to severe gale 9 imminent and storm 10 soon'; this means that an increase is expected to force 9 within six hours and a further increase to force 10 during the following six hours. If the forecaster has been overtaken by events he may have to use the phrase 'gale now increased severe gale force 9'.

With changes in direction, similarly, a warning is issued to indicate a significant change. In this case the convention is that winds are forecast using the eight point compass and that each direction covers 45 degrees either side of a cardinal point. For example 'west' includes any direction from south-west to north-west although the forecaster will, in the first instance, have tried to give the most appropriate average direction. An unexpected change in the wind which takes it outside that sector qualifies for a warning in the form 'west gale 8 veering north-west soon' which means that a gale previously expected to have a direction in the sector south-west to north-west is now expected to veer so as to be in the sector west to north during the period six–twelve hours after the issue of the warning message.

Changes in speed and direction can, of course, be indicated in a combined warning of the form 'west gale 8 backing south and increasing severe gale 9 imminent', or 'west gale 8 backing south imminent and increasing severe gale 9 soon'. In the second example the direction change is expected before the increase in speed while in the first case both speed and direction are expected to occur within the first six hours, if not necessarily at the same time.

Gale cancellation

While gale warnings are forecasts and are essentially a matter of opinion on the part of the forecaster, a gale cancellation is a statement of fact and is only issued when the forecaster is certain, beyond all reasonable doubt, that there is no longer a gale occurring or likely to occur within the area concerned. Although gale warnings may be issued which state that a severe gale 9 is expected to decrease to gale 8 a warning will not be issued to indicate a decrease to below gale force. The shipping forecast, itself, may indicate such a trend but the cancellation is only issued after the gale has ceased and not in anticipation of its ceasing. Gale cancellation messages are broadcast by the coastal radio stations but not by BBC Radio 4 and so the first that the yachtsman knows about a cancellation may be when he hears the next shipping forecast. All gale messages are, however, passed to coastguards and these cancellation procedures should be borne in mind when making enquiries while still ashore. Sometimes a gale warning is cancelled but is expected to be renewed within twelve hours and in this case the message broadcast by the BBC as well as the coastal stations is of the form 'gale ceased but north-west gale 8 expected soon'.

Gale warnings have a lifetime of twenty-four hours and, if still in force at that time after their original issue, have either to be cancelled or 'reissued' as a 'gales continuing' message. These continuation messages are broadcast by the coastal radio stations but not by the BBC; they are also available to coastguards.

Interpreting gale warnings

Many seafarers, both professional and amateur, do not always realise that a gale warning is not a promise that they will definitely experience winds of gale force. A gale warning is simply intended to put the listener on his guard and to point his attention to possible problems. Careful use of the shipping forecast may well indicate that the part of a sea area likely to have the gale is far from where you are sailing. There is little opportunity to issue a warning for part of an area and, in fact, the definition of a gale warning usually makes this unnecessary – a gale warning merely says that gales are possible somewhere in the area. Occasionally, however, a large area, such as

Finisterre or Sole, may have a sharp division with say southerly gales in one half and northerlies in the other. In such a case the forecaster has no option but to issue separate gale warnings for the two parts of that sea area. More usually, luckily, the situation where there are gales from opposing directions in an area is when there is a low pressure area centred over the area with gale force winds circulating around the centre. In this case the gale warning will say 'Cyclonic gale 8' or whatever force is expected. As a matter of general principle, if a gale warning has been issued then it is only prudent to keep a listening watch for any further warnings and on no account miss the next broadcast.

Broadcasting gale warnings

Gale warnings are issued by the Meteorological Office as and when necessary to the BBC and to the Coastal Radio stations run by the PO and the Irish Post Office. The BBC broadcasts the warnings on 1500 metres long wave, 200 kHz, the wavelength which currently carries the national version of Radio 4, at the first programme junction after receipt of the warning. If this junction does not coincide with a news bulletin then the warning will be repeated after the next bulletin. During weekdays the news bulletins are more or less every hour but, unfortunately, this is not the case during evenings or at weekends. As a consequence there may be some considerable time lapse between the issue of the gale warning by the Central Forecast Office at Bracknell and its issue by the BBC. Instead of keeping a listening watch on Radio 4 the yachtsman might well find it more convenient to tune into the coastal radio stations run by the PO. These stations, their frequencies and schedules are listed in *The Admiralty List of Radio Signals, Vol 3*, the RYA booklet *G5*, which deals with weather forecasts, and Meteorological Office Leaflet No. 3 – *Weather Bulletins, Gale Warnings and Services for Shipping*.

The coastal radio stations of the PO (and the Irish Post Office) broadcast weather information for nearby areas – North Foreland Radio, for example, is responsible for Thames, Dover and Wight while Ilfracombe Radio carries information for Lundy and Fastnet. Some of these stations broadcast by R/T in plain language using frequencies in the range 1827 to 2740 kHz, or on VHF, while others broadcast in morse on W/T using frequencies in the range 421 to 472 kHz. Gale warnings are broadcast at the end of the next 'silent period' after receipt of the warning. The silent periods are from 15–18 and

45–48 minutes past each hour for the W/T broadcasts and from 00–03 and 30–33 minutes past each hour for the R/T broadcasts. Warnings will be repeated at the next of the following times 0818, 1218, 1618 and 2018 for the W/T broadcasts and 0303, 0903, 1503 and 2103 for the R/T broadcasts. (All the times of the Coastal Radio broadcasts are in GMT.)

Defining gales and storms

The Beaufort scale defines a gale force 8 as a wind with speeds in the range 34–40 knots but a gale warning will be issued even if the wind is below normal gale limits but gusts are expected with speeds above 43 knots. Similarly a severe gale 9 on the Beaufort scale is for speeds from 41–47 knots but a severe 9 warning may also be issued for gusts above 52 knots. In both shipping forecasts and gale warnings the word 'storm' is used for all speeds of force 10 and greater, storm force 11 and storm force 12 are both terms which you may hear, but with luck only when tucked up in bed at home rather than on the high seas. A warning of force 10 is for speeds in the range of 48–55 knots but also includes gusts above 61 knots.

Storm cones

Originally warnings of gales issued to shipping by the Meteorological Office were by means of *storm cones*, a system instituted by Admiral Fitzroy in 1860 and still in use today. For a time in the early stages there were also *storm cylinders* which were used for force 9 and above. These were soon discontinued and the present system of cones was taken to warn of winds of force 8 or more. Nowadays a cone indicates that a gale is expected within twelve hours, or is already in existence, in the sea area adjacent to the station displaying the signal. A cone hoisted at Hartland, for example, would show that there was a warning of a gale for sea area Lundy. The warning might be in the imminent or the soon categories. The signal is lowered if the wind is less than gale force and if a new warning is not expected to be issued within the next six hours.

Cones are triangular shapes hoisted point upwards or downwards so as to be visible to seaward of the coastguard station wherever the signal is displayed. A *North Cone*, one with the point of the cone upwards, is hoisted for gales from any point to the north of the east-west line. The *South Cone*, point downwards, is for any direction to the south of the east-west line. When the direction is expected to change from the northern to the southern side of the east-west line then the North Cone is lowered and the South Cone hoisted. Conversely for changes in the opposite sense. Unfortunately many of our gales are westerly and the choice of whether to hoist a North Cone or a South Cone may be somewhat arbitrary; the forecaster has to decide if there is any bias to the wind or whether the direction may finally settle down to a direction which would mean that, eventually, he is going to need a North or a South Cone.

Like gale warnings cones have a period of validity, in this case three days, after which they must be renewed if the warning is to continue. Also, like the gale warnings, a cone does not guarantee a gale even in the neighbourhood of the station flying the cone. In fact the wind for which the cone is giving a warning may be at some remote part of the sea area as instanced when the senior forecaster at Bracknell took a telephone call from Portsmouth: 'Do you chaps know what you are doing? There's a South Cone flying outside my window and the wind is light northerly!' A cold front had moved southwards overnight with winds to the south of it reaching gale force from a direction just south of west. To the north of the front the wind was light northerly – hence the telephone call from Portsmouth. At the time of the telephone call the wind over the Channel Isles area was a west-south-west gale force 8 which amply justified the gale warning for sea area Wight and the South Cone at Portsmouth.

At night time the visual signal is a triangular set of lights with the apex of the triangle upwards or downwards in lieu of North or South Cones. Details regarding the locations of lights and cones are found in the appropriate *Sailing Directions*, published by the Admiralty, in the RYA booklet *G5* and in Reeds *Almanac*.

It should be emphasised that the information given by both the visual signals and the broadcast gale warnings should be regarded as only supplementary to the more detailed shipping forecasts broadcast by the BBC and the Coastal Radio Stations.

This may all seem somewhat involved but it is essential that yachtsmen should be fully aware of the meaning of gale and cone warnings, their limitations and methods of distribution.

The shipping forecast

But let us now return to the shipping forecast. It was stated earlier that the shipping forecast is deliberately made invariable in form with only the barest amount of latitude in its construction. In length it has to be capable of being read by an experienced BBC announcer within five minutes starting at 0625, 1355, 1750 and 0015 hours clock time (that is GMT in the winter and BST in the summer). The first three of these broadcasts are followed by the start of other BBC programmes or programme junctions when the various VHF services on Radio 4 join the main programme on 1500 metres. This means that if the shipping forecast has been written in too many words by the meteorologist, as occasionally happens with a forecaster who is new to this particular job, or if the announcer has not had a great deal of experience, then it may not be possible for the whole forecast to be read within the allotted time. Because of the programme schedule the announcer will not normally over-run his time except on rare occasions, and for the late night forecast at 0015 hours. This inflexible time allocation has been a matter for much discussion over the years between the BBC and the Meteorological Office on the one hand and yachtsmen on the other. Although some weather conditions are such that the forecaster would like to be able to extend his forecast by a few minutes it nevertheless remains a fact that the discipline of having to keep to a strict word limit does concentrate the mind to the extent that irrelevancies are omitted. A longer forecast would not necessarily be any better, as it would demand more time from the listener as well as more concentration to ensure that he had missed no important details and would create programme scheduling difficulties for the BBC. Although the enforced brevity of a five minute forecast does create some problems from time to time the current arrangements represent a not too unreasonable compromise which works, on the whole, fairly well.

To be read within five minutes the whole forecast should not greatly exceed 500 words in length. This is because it is read at slightly less than normal script reading speed which is about 110–120 words per minute. The result of this limitation is that the shipping forecast has been pruned to the extent that all redundant words have been eliminated and it is by the position within the forecast as well as by the words used that the listener knows that he is hearing a description of the wind, weather or visibility. *These words, as such, are not used.*

Sea areas

The area covered by the forecast is about one million square miles and is divided into geographical areas known as the *sea areas*. These areas, twenty-nine in number, are of irregular shape and size and have been chosen to have some relevance to the various occupations of professional seafarers. Some are related to fishing grounds, others to the tracks of coastwise shipping and others to the needs of ships on longer passages. The order in which the areas are mentioned in the forecast is invariant, starting with Viking and finishing with South-East Iceland. Two minor exceptions, in a sense, are the forecasts for Trafalgar and the Minches. Sea area Trafalgar (south of Finisterre) only appears in the 0015 forecast, although gale warnings for that area are issued whenever they are needed. Forecasts for the Scottish Fisheries in the Minches, between Butt of Lewis and Cape Wrath in the north and between Barra Head and Ardnamurchan Point in the south, are included after sea area Hebrides in the 0015 and 1355 shipping forecasts from Monday to Friday inclusive.

Whenever possible, areas are grouped together as long as the same general wind pattern applies and the areas are consecutive in the fixed order of the forecast. The pre-ordained choice of the areas does put a constraint upon the forecaster which he could well do without because the weather simply does not recognise the artificial boundaries superimposed by man.

Let us now examine the forecast section by section, see what each part contains and then consider ways and means whereby the listener, seated uncomfortably by his heaving chart table, bobbing up and down in the middle of the North Sea and listening out for the words 'Lee-ho', can record the forecast in sufficient detail for subsequent use.

The preamble

The preamble is an opening sentence which simply states that this shipping forecast was issued by the Meteorological Office at a certain time on a certain day. If you are writing the forecast down as you hear it then, clearly, you do not want to be writing unnecessarily. You will, or should, have already written down the date and the time of the broadcast and so the time of issue need not concern you. The inclusion of this apparently redundant information is simply a safeguard against the announcer reading out either the previous forecast on the same day or even the forecast for the same time from the previous day – after all in typescript one shipping forecast looks much like another. If you tape record the forecast then this heading will enable you to be sure that you are listening to the right one on play back.

Gale warnings

After the preamble the announcer then reads a list of all the areas for which there is a gale warning in force. As with the sea areas later in the forecast he keeps to the standard order. For example, 'There are warnings of gales in Viking, Forties, Cromarty, Forth and Tyne' if there are, indeed, warnings in these areas but 'There are warnings of gales in Cromarty, Tyne, Faeroes and South-East Iceland', if those are the appropriate areas. Gale warnings in all areas are signified by 'There are gale warnings in all sea areas'; if there are four or less areas without a warning then the form is 'There are warnings of gales in all areas except . . .'. The summary does not differentiate between gale 8, severe gale 9, storm 10 and above nor between gales which are imminent, soon or later. If there are no warnings then this is not stated, the forecast simply goes on to the next section of the forecast which is the synopsis.

The synopsis

The *synopsis*, without doubt, is the key which unlocks the whole forecast as it contains a short statement on the lows and highs which are expected to determine the weather over the sea areas during the next twenty-four hours. The synopsis gives the position of these features, their central pressures and their forecast movements. The synopsis also contains information on fronts and troughs if the forecaster thinks that this information is essential and if he has sufficient words available within his overall time limit. The time to which the synopsis refers is up to seven hours before the time of the broadcast, a fact which requires some explanation.

Essentially the reason for this is that meteorological observations are interchanged on a worldwide basis twenty-four hours a day. Among these twenty-four hours there are four main hours when the interchange of data is at its greatest; these times are 0600, 1200, 1800 and midnight GMT. It is at these times also that many merchant ships on passage observe the weather and pass their observations to shore radio stations for transmission to the global meteorological telecommunications circuits, which are capable of passing large volumes of data around the world at high speed. However, despite these rapid channels of communication, it still takes up to three hours after the observations were made before the forecasters at Bracknell have analysed the weather over the North Atlantic/European/American sector of the Northern hemisphere. Additionally, there are always data, especially from ships because of the relatively slow ship-to-shore communications, which arrive up to five or six hours after the data time. This means that the forecasters are continually revising or adjusting their charts during this period. In the shipping forecast the synopsis is intended to be as accurate as possible and, for the various broadcasts, this means that the main information that the forecaster is using is between five and eight hours old by the time the forecast is heard at sea. This is not to say that the forecaster does not use any more recent information; he does, but it is by no means as complete as that for the 'main' hours listed and he can by no means be as certain about the positions of the various weather features at any intermediate time as he is at the four main synoptic hours. The result of all this is that the forecast broadcast on 0015 has a synopsis time of 1800 when working on GMT or 1900 when on BST. The forecast at 0625 has a synopsis time of midnight GMT or 0100 BST, that at 1355 a synopsis time of 0600 GMT or 0700 BST and the 1750 broadcast a synopsis time of 1200 GMT or 1300 BST.

Whenever centres of pressure are mentioned in the synopsis the order of words is 'low' or 'high', as the case may be, then the position at the time of the synopsis, followed by the central pressure (in whole millibars). The forecaster can then indicate how he expects the system to move over the next twenty-four hours in one of two ways. He can either give the expected position twenty-four hours hence and a new, forecast, central pressure value or, alternatively, he can give a direction of movement using the eight point compass and a speed. He indicates the *speed* by one of the following adverbs:

slowly meaning 0–15 knots
steadily meaning 15–25 knots
rather quickly meaning 25–35 knots
rapidly meaning 35–45 knots and
very rapidly meaning over 45 knots.

There are two points to note here. First the direction is only given to the accuracy of the eight point compass, to do better would be to give an impression of precision which is not possible and the same applies to the speeds which are given using the coded words mentioned. The second is that words like 'moving north-east rapidly' are less likely to be misunderstood in conditions of poor radio reception than, say, 'moving 45 degrees at 40 knots'. The descriptive terms and the use of the eight point compass, therefore, not only give realism to the forecast but are also aids to clarity.

The reference to forecast positions of the lows and highs in twenty-four hours' time *after the synopsis time* is because the British Meteorological Office, as its standard practice, produces forecast charts for times in multiples of twenty-four hours after the data time. Thus the forecaster writing the shipping forecast has an actual chart for one of the times 0600, 1200, 1800 or midnight GMT, that is a chart of actual weather analysed to show as precisely as possible the positions of weather systems at these times, and a forecast chart showing the expected positions, shapes and pressure values of the same system at 0600s, etc., the following day. As the forecast is valid for twenty-four hours from the time of broadcast, that is a further five or more hours beyond the time for the future positions of the synoptic features described, the discerning listener may find some apparent discrepancies between the forecast winds for the sea areas and the forecast positions of the highs and lows.

In the synopsis, when talking about centres of pressure, the forecaster may refer to a *deepening* or *filling low*; these terms arise from the analogy drawn in Chapter 1 between the atmosphere and a bowl of syrup. A deepening low is one in which the central pressure is becoming lower and, therefore, more active while a filling low is one in which the central pressure is rising and the low becoming less active. The reasons for this are that when the central pressure in a low is decreasing it almost certainly follows that the pressure gradients around the low are becoming stronger and the winds are increasing. The converse is true for the filling low.

A *building high* is one in which the central pressure is increasing while a *declining high* is where the central pressure is decreasing – like a mound in the syrup bowl settling down back to the general level of the syrup. A building high is one which may be developing into a blocking high while a declining, or weakening, high is one which may have been blocking but is now giving way and allowing the lows to make eastwards progress rather than be diverted to the north or south by the high.

A *low* which is described as *complex* is one which has two or more identifiable centres. The behaviour of these complex lows is often difficult to describe, first, because the separate centres are not always known very precisely from the observations, at least when the low is still over the sea, and, secondly, because, the separate centres do not always retain their identities for as long as twenty-four hours. While they do exist, these complex centres tend to move around each other, in a dumb-bell fashion, circulating with the wind flow of the low (that is, anti-clockwise in the northern hemisphere).

When the forecaster mentions *troughs* or *fronts* in the synopsis he will again give positions and movements in the same form as for the pressure centres but, this time, there are no central values to worry about.

Positions of weather systems are given by means of reference to well known geographical features rather than by the use of latitude and longitude; as in the case of the speeds and directions of movement this is to avoid the use of numbers which might well prove difficult to distinguish against a background of interference. The geographical features used for this purpose include the sea areas themselves and well known parts of the United Kingdom or other countries. Because this forecast is written, in the first instance, for commercial shipping the forecasters feel free to refer to such places as Cape Farewell (the southern tip of Greenland), the Denmark Strait (between Greenland and Iceland), the Davis Strait (between Greenland and Canada), the North Cape (northern tip of Norway) and the Norwegian Sea (between Iceland and Norway). Apart from these the locations referred to will be, or should be, familiar to all sailors.

Forecasts for the next twenty-four hours

After the synopsis come the forecasts for the sea areas for the next twenty-four hours, that is twenty-four hours from the time of the broadcast. Each sea area is given in the strict order referred to earlier and any grouping together of areas is only permissible if they are consecutive within the fixed order. The forecast for each area is in the order: wind–weather–visibility, although these words are *not* used.

The wind is described using the eight point compass and the Beaufort scale; as with the movement of pressure systems to use degrees and knots would imply a greater precision than is justified and would, in any case, lead to a loss of clarity. Because the eight point compass directions imply a range of directions of ± 45 degrees and because the Beaufort scale implies a range of speeds there is, already built into the terminology, a recognition of the imprecise nature not only of the forecasts but also of the weather itself. When describing the expected changes in the wind the forecaster uses the ordinary words of the English language in their dictionary senses.

These may include: increasing, decreasing, becoming, veering (changing direction in a clockwise sense e.g. from west to north-west or east to south), backing (changing direction in an anti-clockwise sense e.g. from north to west).

The weather is usually given as 'rain', 'showers', 'drizzle', 'thunder' or combinations of these. 'Fair' is used to mean that little of any significance is expected in the way of precipitation, mist or fog; it does *not* carry any implications regarding the amount of cloud or sunshine. The weather section may well include qualifying words such as 'at times', 'in places', 'at first', 'later', 'locally', 'spreading', 'here and there', 'heavy', 'then' – in the sense of 'rain then showers' – 'early' and so on. All these words have their obvious meanings.

The visibility is given last in the form of 'code' words. These are 'good', meaning more than 5 miles; 'moderate', meaning 2–5 miles; 'poor', meaning 1100 yards to 2 miles; and 'fog' meaning less than 1100 yards. As with the weather, variability in time and space and expected changes or trends are described using the ordinary words of the English language. The description of the visibility is an example of meteorological terminology being varied to suit the user. In land area forecasts the word 'fog' is used to describe visibilities of less than 200 yards. A motorist would hardly regard 400 yards, for example, as a fog but the master of an oil tanker would and so would the captain of an airliner coming in to land. If the visibility in the shipping forecast is 'fog' then the weather may be left blank. Fairly obviously, and anyone who tries to write down a shipping forecast can demonstrate this for himself, it is extremely difficult, if not impossible, to compress all the weather expected over the sea areas into the broadcast time allowed, even assuming that there are no uncertainties. Thus the forecaster is forced to take a very simplistic view of the wind and weather developments using a broad brush type of approach in his writing of the forecast. He simply has no option but to group the sea areas together as far as possible using as his main criterion his ability to describe the wind in the same general manner for the areas so grouped. This may mean that he has to over-simplify the weather description for these areas or, sometimes, use a device such as 'but in east Sole at first' for some weather element.

On occasions the forecaster has to split an area in half, sea area Finisterre, for example, is a large and often unwieldy area and is quite often split into northern and southern parts. In most cases, however, the weather is moving across the areas, usually from west to east, and the problem is to describe a wind change, probably accompanied by changes in weather and visibility, moving across a number of areas. If possible the forecaster will try to give some idea of the timing of such changes using words like 'at first', 'early' or 'later'. Occasionally you may hear the other timing words used in gale warnings: 'imminent' and 'soon' which will be used in their gale warning sense. More usually, however, the forecaster has to assume that the listener can relate the changes he is describing to the information in the synopsis and the sea area forecasts will simply use words like 'becoming' or 'then'.

Icing in South-East Iceland

The sea area forecasts finish with a statement relating to *icing in South-East Iceland*. This information is for the fishing fleets in those waters and is a reminder that the shipping forecast has a large professional audience to whose safety and efficiency it forms a significant contribution. The forecast of icing conditions is because of the rapid accretion of ice that can occur on the superstructure of trawlers in certain very critical conditions of wind, sea temperature and air temperature; this icing can cause the vessel to capsize.

Weather reports from coastal stations

Following the sea area forecasts come the weather reports from the coastal stations. These are given for twelve or thirteen stations around the coast, the number depending upon the time of day and availability of time. Like the rest of the forecast the reports are always read in the same order starting with Tiree in the Hebrides and working clockwise around the north of Scotland, down the east coast of England, along the south coast and up the west of the British Isles to Malin Head on the northern coast of Ireland. Each report is given in a fixed order: wind – weather – visibility – pressure – pressure change.

Wind is given in the 32 point compass and Beaufort force. The 32 point compass is N, N by E, NNE, NE by N, NE, NE by E, ENE, E by N, E etc. This may seem laboured but, again, is in the interests of clarity when heard through radio interference.

The weather is given as in the forecast but differentiation is now made for intermittent or continuous precipitation, light, moderate or heavy precipitation, and the weather in the last hour may also be mentioned, for example, 'recent rain' may be very significant.

There is also scope for the report to mention precipitation within sight. Haze or smoke may be given as the reason for a poor visibility rather than mist or fog. This, again, is of some importance because with smoke or haze (dust) it is unlikely that the visibility will deteriorate further while with mist a little more cooling would give some further condensation which would cause the visibility to worsen. In the coastal reports 'fog' is still used for visibilities below 1100 yards but 'mist' is used to describe visibilities between 1100 and 2200 yards unless it is due to smoke or dust haze or for some other (obvious) reason such as drizzle, snow or heavy rain.

The barometric pressure is reported in whole millibars and the recent change in the pressure is indicated by means of descriptive terms which have coded meanings. If, over the past three hours, there has been a general rise or fall in the pressure then this is given as 'rising' or 'falling' qualified by 'slowly' meaning a change of 0.1 to 1.5 millibars in the period, 'quickly' meaning a change of 3.6 to 6.0 or 'very rapidly' meaning a change of more than 6 millibars. No qualification implies a change of 1.6 to 3.5 millibars. 'Steady' means a change in pressure of less than 0.1 millibars in the three hours.

Other terms are 'now falling' which means that the barometer has been rising but is now falling, that is, it has just recently begun to fall. Similarly, 'now rising' means that, after a fall, the pressure has just begun to rise. The term 'falling more slowly' means that after falling the pressure is now steady or that the pressure is falling much less markedly. Conversely, 'rising more slowly' means that a rising barometer has steadied or is showing signs of steadying.

The times of the coastal reports are for an hour and a half or two hours prior to the time of the broadcast. This is because all the stations observe the weather just before each hour and immediately pass their reports to meteorological collecting centres for transmission to Bracknell; in the case of the Dutch Light Vessel Noordhinder this is via Amsterdam. The reports are received at Bracknell in the international number code form used worldwide for the interchange of weather information; from this code they have to be translated into the plain language form used in the shipping forecast. The time taken for these various processes means that the latest reports which can be included in the bulletins must be at least one and a half hours old by the time of the broadcast. Sometimes there are delays in transmission time between the reporting station and Bracknell and in this case the previous report will be used if this is not more than three hours old. The text will include a mention of the time of any report from the previous hour.

The coastal stations which are used vary from time to time for a number of reasons. For example reports become no longer available, perhaps a light vessel has been replaced by a buoy; a land station, perhaps a coastguard station, is no longer manned. Sometimes it becomes possible to replace one report by a better, more representative one. In recent years, at various times, Wick has been replaced by Sule Skerry,* Prestwick by Malin Head, Galloper by Noordhinder and Portland Bill by Channel Light Vessel. Also, the report from Jersey is now included whenever there is time within the broadcast schedule. There are occasions when, for some reason or other, a station may be unavailable temporarily and in this case a short term substitute may be used but which may not be as useful to the mariner as the normal report. It is hoped that an unrepresentative report is better than no report at all.

Conclusion

The shipping forecast is, thus, a forecast finely tuned to the needs of the user but is subject to change, partly because of force of circumstances and partly because of periodic reviews by a Department of Trade and Industry (formerly Board of Trade) committee which is responsible for safety of navigation. This committee includes representatives from the Royal Navy, the Ship Owners, the Lighthouse and Coastguard organisations, representatives of the various seamen's trade unions or staff associations, the Harbour Authorities, the Post Office and the Royal Yachting Association. The Marine Superintendent of the Meteorological Office, himself a master mariner, acts as adviser to this committee on meteorological matters. The composition of the committee emphasises that the shipping forecast has to meet a very wide range of needs and that the majority interest is the professional one. The Meteorological Office provides the forecast in the manner determined by the user, in so far as these requirements can be met. The roles of the BBC and the PO are simply to act as media by which the forecast is passed to the mariner from the forecaster. In the case of the BBC this tends to be taken for granted but it should be remembered that there is nothing in the charter of the

*Sule Skerry has now been replaced by Sumburgh in Shetland.

BBC to say that shipping forecasts must be broadcast. The BBC has many functions, many other people to serve and many other priorities; these are facts which are not always recognised by users of the shipping forecast.

Because of the confusion in the minds of many users of the shipping forecast regarding the various times in the different parts of the forecast these have been summarised in Table 3, p. 102.

6

Recording, Analysing and Using the Shipping Forecast

Having described the content of the shipping forecast it should be apparent that a wealth of information is available to the sailor and we shall now go on to the next most important step of seeing just how this can be used in a systematic manner.

Recording

The first stage is to ensure that we have a sufficiently good record of the forecast so that there is no uncertainty regarding what was said. 'Did he say south-west or was it north-west? What was the wind in Sole later?', and so on. In these days of battery-operated tape recorders many yachtsmen find it a convenient practice to record every forecast even when the weather seems set fair. The forecasts can then be re-played as often as necessary so that all doubts can be resolved. Unfortunately, tape recorders, like many electro-mechanical devices, are subject to failure and can almost be guaranteed to fail in the most difficult weather just when you really do need the forecast. Additionally, the systematic use of the forecast does require its being written down in some form sooner or later so that it can be studied as a whole and the overall pattern can be seen. It is, therefore, worthwhile developing the habit of taking the forecast down on paper as it is read out and only using the tape recorder as a back up system for the occasional words which can still get missed even with a good deal of practice. After a while, and using the techniques described in this chapter, it should be possible to take the whole forecast down as it is read using the tape recorder only for the few words which got missed when someone called 'Gybe-Oh!' or 'Tea up!' just at a critical moment.

On first hearing a shipping forecast most people despair of ever being able to take down more than a few words, even using short-hand methods. But it is possible – even in a blow! The Meteorological Office publishes pads which are available from weather centres and which consist of maps of the sea areas around the United Kingdom with lines printed over the areas so that each area forecast can be written down over the areas in question. This is quite useful for looking at the forecasts for areas adjoining the ones in which you are interested but it is not designed to allow you to go further and attempt to draw your own isobars and fronts and so maximise the benefit that can be obtained from the forecast. For this purpose you can buy the pads of maps, published jointly by the RYA and the Royal Meteorological Society, which consist of maps of the sea areas with a table on the reverse side (see Fig. 11, p. 58.). The table is designed to help you to take down the forecast with the maximum efficiency and is shown in Fig. 16, p. 110.

The procedure for taking down the forecast will now be described step by step. First of all and before the forecast comes on the air you should write down the date and time of the forecast and the times for both the synopsis and the coastal reports; for these you can refer to Table 3. If the forecast is preceded by a gale warning summary then simply run your pencil down the left hand side of the list of sea areas and tick any area for which there is a warning in force. Remember that the warnings are given in the same order as on the form so that each area mentioned will be lower down the sheet than the previous one although, obviously, not necessarily consecutively. This noting of the gale warning areas is not strictly necessary but is simply a check which might be useful later giving, as it does, some preliminary warning regarding areas which may be of particular interest.

After the gale warning summary if there is one, or immediately after the preamble otherwise, the announcer says, 'The general synopsis at . . .'. As you should have already written down this time these words allow you to position your pencil on the part of the form for the synopsis. If the announcer then says 'high' or 'low' simply write down H or L. The position of the centre which then follows is more difficult to record but by the use of initials for the geographic locations and for the directions of the centres from these locations it is possible to take down brief notes in the form 'L 600 W Ir 78' for 'Low 600 miles west of Ireland 978' or 'L N Ro 92' for 'Low north Rockall 992'. Note that the central pressures will be in the range from 920 millibars for a very deep low to 1050 for the middle of a very intense high. There is thus no ambiguity in missing out the 9 or the 10 at the beginning of the pressure; the

Broadcast times		0015	0625	1355	1750
Forecast valid to		same time next day			
Times of synopsis	GMT BST	1800 previous 1900 day	midnight 0100	0600 0700	noon 1300
Times of forecast positions	GMT BST	1800 today 1900	midnight 0100	0600 0700 'tomorrow' (Or 'By same time tomorrow')	noon 1300
Coastal reports		2300	0500	1200	1600

Table 3   Summary of times used or referred to in the shipping forecast. Note that the broadcast times and the times to which the coastal reports refer are all 'clock time' and, so, are invariant from summer to winter. The times in the synopsis do, however, change from summer to winter depending upon whether the time standard is GMT or BST.

These times are correct at time of going to press but are subject to change.

two digits are quite sufficient. As far as the positions of the centres are concerned you should develop your own abbreviations so that, at the end of the forecast, you can go back and fill in the necessary letters only having to remember what that funny little squiggle meant for a very few minutes. But do this immediately; an hour or so later not only will the forecast be that much older but you may already have forgotten just what some of your abbreviations meant. At first it is suggested that you omit such words as 'complex' or 'weak', you might remember to add these afterwards but, in any case, as you get more practice you will find that these, or rather some appropriate abbreviations, can be included without too much difficulty.

Turning now to the movements of highs and lows, if the positions at twenty-four hours after the synopsis time are given then just take these second positions down. Do not bother noting the times mentioned – it can only be twenty-four hours later than the synopsis time. Similarly, if the low is deepening or filling then the forecaster will give a second pressure, so just take this down and do not bother with such words as 'filling'; this will be obvious from the values of pressure. The same applies for a high. With experience words like 'filling' provide a check that the second pressure should, in this case, be higher than the first. If the forecaster gives the movement of a low or a high by means of a direction and a speed word such as 'slowly' or 'rather quickly' then these can be written as 'NE sl' for 'moving north-east slowly', or 'E rq' for 'moving east rather quickly'. Alternatively the direction can be written as an arrow: → for 'east' or ↗ for 'north-east' etc.

Information about fronts or troughs, when given, is taken down in a similar manner. Unfortunately the terminology is somewhat variable; you will hear mention of either fronts, cold fronts, warm fronts, occlusions, frontal troughs or troughs. These can be written as F, CF, WF, O, FT or T respectively. One of the problems is that a frontal system may well have not only a cold front, warm front and occlusion but that the occlusion process will be proceeding during the period of the forecast. There is simply not enough time available to describe such systems and the forecaster may have no option but to use the omnibus word 'troughs'. Also there may be a cold front followed by troughs in the isobars which are not associated with any change in air mass and these latter troughs may be more significant in terms of a change in the wind than the cold front itself. There will be occasions when the forecaster simply cannot mention even one front or trough because of the overall length of the forecast or he may be able to mention, perhaps, one trough or front and then has to decide which is the most significant.

If positions are given it may be something fairly simple like 'Approaching Bailey', which can be written 'app Ba', or in terms of a number of sea areas or parts of sea areas, for example 'Cold front east Faeroes to Cromarty to Cornwall' could be taken down as 'CF E Fa – Cr – Cor'. To take that much down is as much as you can hope for while desperately keeping your fingers crossed that you will know what it all means five minutes later! Movements of troughs may also be by either direction and speed or by a second set of positions. The same advice applies as for the information relating to pressure centres – simply take down as much as you can, leave anything that you find too difficult, and hope that you can fill in the details after the broadcast has finished. At the end of the synopsis the announcer always says the words. 'And now the area forecasts for the next twenty-four hours', this gives you about five seconds to catch up with the end of the synopsis before going on to the next section of the shipping forecast.

The area forecasts will be read out as single areas or as groups of areas which are consecutive on the form and so if you place your pencil by the right hand side of the word 'Viking' then as the announcer says 'Viking, Forties, Cromarty' (or whatever) you can simply run your pencil down the column marking all the area names as he reads them. After the first list of names the announcer then gives the wind for those areas. The direction is given by means of the eight point compass and the speed on the Beaufort scale. Speed and direction can both vary and this can be expressed in several ways. Typical examples might be 'South 4 increasing 5 or 6', 'South west 4 or 5 veering north-west increasing 7 later', 'South 5 or 6 becoming cyclonic variable then north-west 7 perhaps gale 8 later', 'West gale 8 or severe gale 9 decreasing 5 to 7'. By means of suitable abbreviations and writing these down in sequence under the column headed 'Wind' it is reasonably easy to take them all down. Possible abbreviations for these various forecasts could be: 'S 4 → 5/6', 'SW 4/5 → NW → 7l', 'S 5/6 → CV → NW 7 p8l', 'W 8/9 → 5/7'. The oblique sign / can be used to indicate a range of speeds or directions and an arrow → can be used to mean a change; it does not matter whether this is veering, backing, increasing or decreasing because this should be evident from the sense. If, of course, you can take down the actual word, or an understandable abbreviation, then so much the better because this will act as a quality control on the rest of that section of the forecast. 'l' for later and 'p' for perhaps are fairly obvious. CV or $\mathfrak{H}$ could be used for cyclonic variable. Words such as gale, severe gale and storm, which are always used with forces 8, 9 and 10–12, need not be taken down, the number of the force is quite sufficient and the word 'gale' and so on merely draws your attention to the force and increases the possibility of you taking down the force correctly.

After the wind the weather is written in the column headed 'Weather' with single letters used for the various possible elements that are likely to be mentioned. Examples would be 'r' for rain, 'd' for drizzle, 's' for showers, 't' for thunder and so on. Snow would probably have to be 'sn'. If the weather is given as 'fair' or 'mainly fair' then this can be left blank, with no ambiguity. Changes from one type of weather to another can again be shown by means of a dash or an arrow. Words such as 'locally', 'at times', 'here and there', 'occasional' can all be written as 'loc', 'occ', or possibly by means of brackets. Some examples are 'Fair early, rain later', 'Showers spreading east', 'Rain then showers', 'Rain at times', 'Rain in east at first then occasional showers'. Possible versions of these are respectively 'r l', 's → E', 'r → s', 'r loc' or 'r occ' or '(r)', and 'r E → (s)' or 'r E → occ s'. In the majority of cases it will hardly matter if you omit such words as occasional or locally, the important words are those which describe the type of precipitation because it is this which give the indication of the type of air mass. Sometimes the forecaster may say heavy rain or heavy showers and this can easily be recorded by use of capital letters, 'R' for heavy rain for example.

Some people find it easier to use symbols or a mixture of letters and symbols such as those that are used on weather charts, that is ✳ ● ꝰ and ▽ for snow, rain, drizzle and showers respectively. The best advice is to experiment and see whether you find symbols or letters easier to use. One notable advantage of the symbols is the obvious discrimination between showers and snow.

Visibility is reasonably easy to write down as the only descriptions used are 'good', 'moderate', 'poor' and 'fog' and the obvious abbreviations are g, m, p and f. The variability and trends in visibility may be expressed in precisely the same way as for the wind and the weather. Fog banks, a phrase used to describe dense areas of fog, could be written as 'fb'. The fog symbol is ≡.

The important point about all these abbreviations is that you should use symbols or letters which *you* will understand. They are for your use and there is no right or wrong way to take down the forecast, there is only the requirement that you should be able to know, reasonably accurately, what the announcer said even if you have to fill in gaps or expand some of the cryptic notes that you made after the broadcast.

The final part of the broadcast, the reports of actual weather which have been decoded from routine meteorological observations, may contain rather more data than the sailor can use. Nevertheless, it is good practice to take them down as fully as possible, if only for completeness and because you never know just which bit of information you may or may not want to use later.

For the coastal reports the wind should be taken down in the obvious way, for example SW 4, WNW 6 and W/S 3, the last one being west by south as the 32 point compass is used. The weather can be very varied as there are up to 100 ways, using the international weather codes, in which a weather observer can report the actual weather although not all of these are broadcast in the shipping forecast. Some of those that are broadcast refer to rain, drizzle or snow all of which may be light, moderate or heavy, qualified by either intermittent or continuous. Rain, drizzle, snow, showers or fog may all be reported as being 'in the last hour' which implies 'but not at the time of the observation'. Showers may be moderate, heavy or not qualified by any such term and may be of rain or snow. Precipitation may be within sight reaching or not reaching the ground. These may be mist, smoke haze or just haze. There may be thunderstorms with hail, or snow or squalls. Rain and snow may be mixed and so on. If there is no 'significant' weather then the weather is not mentioned and the announcer goes on to the next part of the report.

How should you get these down? For the various kinds of precipitation the letters r, d, and so on can be used to mean light or moderate rain or drizzle, capital letters R, D can be used for heavy rain or drizzle, repeated letters can mean continuous precipitation while single letters mean intermittent. 'RR' would be continuous heavy rain and 'd' would be light or moderate intermittent drizzle. A single bracket could be 'in the last hour', 's)' would be shower in the last hour. The shipping forecast terminology is 'recent showers'.

This 'recent' observation can be a very important piece of information, particularly when it refers to rain, drizzle or fog because it often indicates that there has been a change of weather type, perhaps from a warm sector to the cold air behind the warm front or from the rain ahead of the warm front to the warm sector. A shower in the last hour is not usually very significant as showers are short-lived, transient features and the 'recent' report simply increases the chance of showers being reported and so is useful from that point of view. Sometimes, however, the observer will report 'shower in the past hour' when a weak cold front has passed in which case all he might have seen would be a short burst of rain from convective type cloud. 'Precipitation within sight' usually means showers and could be written as such or could be written using brackets, e.g., (r), the word 'precipitation' is used in this case because the observer does not know whether he is looking at rain, hail or snow.

The symbols for rain, drizzle and other weather phenomena are convenient for use when taking down the reported weather because by doubling up the symbols horizontally or vertically it is possible to express continuous for the former and moderate or heavy for the latter. •, ••, ⁚, ∴, ⁝, ⁘, would be intermittent light rain, continuous light rain, intermittent moderate rain, continuous moderate rain, intermittent heavy rain and continuous heavy rain respectively. Similarly for drizzle and snow. Rain or snow showers can be $\overset{\bullet}{\nabla}$ or $\overset{*}{\nabla}$. A hail shower can be $\overset{\triangle}{\nabla}$, a thunderstorm can be ☈ Fog and mist can be ≡ or = while haze can be a 'z' or ʃ∼ (a smoking chimney). Fog may be reported as thick but this need not be recorded as the visibility report will show this. Similarly thunderstorms may be with heavy showers but this need not be recorded, the fact that there are thunderstorms is sufficient indication that the showers might be heavy. If you want to show heavy showers then, perhaps, a line above the shower symbol will do: $\overline{\nabla}$.

Some of the detail may seem rather like gilding the lily and if you want to keep to items of essential interest only, then try to note somehow whether or not there is rain or snow, drizzle, showers or thunder, mist or fog, or whether any of these has been reported as occurring now or within the last hour. With a little practice you will probably find that it becomes possible to take down the extra information and, in any case, with experience you will soon find the implications of the different descriptions and come to know those which are worth recording.

After the weather, if reported, comes the visibility. This is given as a number of yards or miles. A number like two or ten can only mean miles while one like 500 or 2000 can only be yards so there is no need to write down which it is as omission of the 'miles' or 'yards' does not lead to any ambiguity. Sometimes the report will say 'more than 32 miles', or some such number. All this means is that the observer is judging the visibility from objects which he can see and that, in this case, his furthest point is at 32 miles and yet he is sure that the visibility is better than that. Such values are academic to the sailor and the best thing to do is to call this 32 miles and just write down 32.

After the visibility the last item in the report is the barometric pressure which is given in whole millibars with a description of the behaviour of the barometer within the last three hours. As with the synopsis there is no need to write the pressure down completely, the last two figures will do. In the part of the shipping forecast form used for writing down the coastal reports the last box in each line is headed 'Change' and this box can be used very simply to indicate falling or rising, qualified by the various adverbs mentioned and defined in the last chapter. Falling or rising not qualified by any adverb can be shown by a line corner to corner upwards (left to right) for rising and downwards for falling. 'Slowly' can be a less steep line; a steeper line can represent 'quickly'; and a steeper line still can be 'very rapidly'. A horizontal line can, rather obviously, be steady. The other possibilities, 'now falling', 'rising more slowly', 'now rising' and 'falling more slowly' can be shown as ∧, ∕‾ , ∨ or ＼_ respectively. The last two of these are probably the most significant of all as they often show that a trough or a front has just passed (now rising) or is just passing (falling more slowly) the reporting station.

The following is a shipping forecast script:

And now here is the shipping forecast issued by the Meteorological Office at 1700 hours on the twenty-sixth of May.

There are warnings of gales in Hebrides and Bailey.

The general synopsis at 1300. Low 300 miles west of Bailey 984 expected north Faeroes 976 by 1300 tomorrow. Associated trough Rockall to Shannon to west Sole moving steadily east.

And now the area forecasts for the next twenty-four hours.

Viking Forties. West 4 or 5 backing south and increasing 6. Rain later. Good becoming poor.

Cromarty Forth Tyne. South-west 4 or 5 veering west later. Rain at times. Moderate or poor becoming good later.

Dogger Fisher German Bight. South-west 3 or 4 backing south and increasing 5. Rain later. Moderate becoming poor with fog patches.

Humber Thames Dover Wight. Variable or south 3 becoming

south-west 4. Rain or drizzle. Moderate becoming poor with fog but good in Wight later.

Portland Plymouth Biscay. South-west 3 becoming north-west 5 or 6. Rain or drizzle then showers. Poor with fog becoming good.

Finisterre. South-west 3 becoming north-west 4. Mainly fair. Poor with fog becoming good.

Sole Lundy Fastnet Irish Sea. South-west 5 becoming north-west 6 or 7. Rain or drizzle then showers. Poor with fog becoming good.

Shannon Rockall. South 7 in east at first otherwise west to north-west 5 or 6. Showers. Mainly good.

Malin Hebrides Bailey. South to south-east 5 or 6 but gale 8 in Hebrides and Bailey at first becoming west 6. Rain or showers. Moderate becoming good.

Fair Isle Faeroes. South to south-east 4 increasing 6 veering west later. Rain spreading north. Moderate becoming poor.

South-East Iceland. South to south-east 4 increasing 6 but cyclonic variable later. Rain spreading north. Moderate or good becoming poor.

No icing in South-East Iceland.

And now the weather reports from coastal stations for 1600.

Tiree. South-south-east 4. Continuous moderate rain. 3 miles 1010. Falling.

Sule Skerry. South 3. 10 miles. 1012. Falling slowly.

Bell Rock. South-east 2. 10 miles. 1015. Falling slowly.

Dowsing. South 2. 15 miles. 1020.

Noordhinder. Calm. 21 miles. 1021. Falling slowly.

Varne. Calm. 15 miles. 1021.

Royal Sovereign. South 1. 7 miles. 1021. Falling slowly.

Channel Light Vessel. South-west 3. Mist. 2000 yards. 1020. Falling.

Scilly. South-west by south 3. Fog. 550 yards. 1018. Falling slowly.

Valentia. South by west 4. Rain and drizzle. 2000 yards. 1010. Falling.

Ronaldsway. South-south-west 3. Continuous moderate rain. 2 miles. 1015. Falling.

Malin Head. South 3. Continuous moderate rain. 3 miles. 1010. Falling more slowly.

That is the end of the shipping forecast.

Analysis

A version of what might have been taken down is shown in Fig. 16.
By itself the very fact that the forecast has been recorded, if not
for posterity but for immediate use at least, is of some value simply
for ease of reference. There are, however, a number of questions
that the sailor using this forecast might wish to ask. For example,
the visibility in sea area Wight is expected to be poor but is also
expected to improve later – the question is when? You might be
on a Channel crossing hoping to make for Lymington and are
anxious about your landfall on the English coast. The wind in the
Channel, sea area Plymouth, say, is forecast to veer and increase
but, again, when? As the forecast stands there is no obvious way
in which these kinds of question might be answered. What if you
were in sea area Fastnet to the south of Ireland and the wind veered,
as expected from the forecast, with an improvement in the visibility
but the pressure then began to fall after a short rise – what would
that mean? Or you might be at anchor in the Fowey River and
wishing to make a passage across the Bay of Biscay – is it worthwhile
leaving now or later?

To be able to answer these and other questions necessitates a little
work constructing a chart which looks something like that used by
the forecaster when writing the forecast. The description of how
to do this may sound rather cumbersome and your first efforts will
certainly be very laborious. The technique is one which you can
practise at home and, with some experience, it should be possible
to produce fairly quickly a chart which will give you some help.
You will not get all the answers right first time but you will do rather
better than if you just try using the forecast literally, for you will
start to develop a better understanding of the weather and its
patterns. If you are racing then you will be able to make decisions
more soundly; if you are cruising then you will be able to reduce
some of your passage times and make more informed decisions about
which harbour to make for and when to leave harbour. If on shore
and speaking to a forecast office on the telephone you will be in
a better position to ask the forecaster relevant questions.

Fig. 16 A transcript of the shipping forecast broadcast at 1750 on
26 May as taken down on the Royal Meteorological Society/Royal
Yachting Association Metmap pad. NB: Sumburgh, in Shetland, now
replaces Sule Skerry.

R.MET.SOC./R.Y.A. METMAP

GENERAL SYNOPSIS at 1300 GMT/BST 26 May

L 300 W Ba 84 →NFae 76

T Ro - sh - W So. st E

Gales	SEA AREA FORECAST		Wind	Weather	Visibility
	Viking	⌐	W 4/5 → S → 6	• ∫	g → P
	Forties	⌐			
	Cromarty	⌐			
	Forth		SW 4/5 → w ∫	(•)	mP → g∫
	Tyne	⌐			
	Dogger	⌐			
	Fisher		SW 3/4 → S → 5	• ∫	m → P(≡)
	German Bight	⌐			
	Humber	⌐			
	Thames		V/S3 → SW 4	•/ɔ	m → P≡
	Dover				→ g wi∫
	Wight				
	Portland	⌐			
	Plymouth		SW 3 → NW 5/6	•/ɔ → ∇	P≡ → ;
	Biscay				
	Finisterre		SW3 → NW 4		P≡ → g
	Sole				
	Lundy	⌐	SW5 → NW 6/7	•/ɔ → ∇	P≡ → g
	Fastnet				
	Irish Sea				
	Shannon	⌐	S7 in E early /W/NW 5/6	∇	g
	Rockall				
	Malin				
✓	Hebrides		S/SE 5/6 (8 H+B early) → w6	•/∇	m → g
	Minches				
✓	Bailey				
	Fair Isle	⌐	S/SE 4 → 6 → w∫	• ↑	m → P
	Foeroes				
	SE Iceland		S/SE 4 → 6 → ↺∫	• ↑	m/g → o

└ Mark gale areas └ Connect areas grouped in forecast

COASTAL REPORTS at 1600 BST/GMT	Direction	Force	Weather	Visibility	Pressure	Change
Tiree	SSE	4	·•	3	10	
Sule Skerry	S	3		10	12	
Bell Rock	SE	2		10	15	
Dowsing	S	2		15	20	
Noordhinder		O		21	21	
Varne		O		15	21	
Royal Sovereign	S	1		7	21	

COASTAL REPORTS	Direction	Force	Weather	Visibility	Pressure	Change
Channel L.V.	SW	3	=	2000	20	
Scilly	SW/S	3	≡	550	18	
Valentia	S/W	4	ɔ	2000	10	
Ronaldsway	SSW	3	·•	2	15	
Malin Head	S	3	·•	3	10	
Jersey						

It should be emphasised that the time taken to draw a chart should not be excessive, if it takes too long then its value decreases; with practice a total time of 15–20 minutes after the end of the forecast should be the limit and it is worth spending time practising to achieve a reasonable result within that time.

Using the forecast reproduced above we shall now draw a chart of isobars and fronts for 1600 BST. Why choose that time? Because that is the time for which the coastal stations reported their weather observations and we shall be trying to draw isobars to the reported pressures. It will be necessary, therefore, to update some of the other 'actual' weather information, namely that contained in the synopsis, to that same time. Before this can be done it is necessary to present the information which we have on to a Metmap. This has been done in Fig. 17 and is fairly self-explanatory.

First the winds, weather and visibilities for the sea areas as expected at the *beginning* of the forecast period were plotted using a mixture of symbols and letters or figures. Winds have been shown using arrows with feathers, the arrows are flying with the direction of the wind and the number of feathers show the wind force. One short feather is one force and one long feather represents two Beaufort forces, so that a force 3 is shown by one long and one short feather, while a force 4 is shown by two long feathers. Examples plotted are the force 3 in Portland, Plymouth and Biscay and the force 4 in Faeroes and Fair Isle. How to represent a range of forces such as the force 4 or 5 expected in Viking and Forties, is not easy to decide. The safest method, probably, is to plot the higher force of two or the average if a range of three forces is given. The expected weather at the beginning of the period has been shown using the symbols described above, although there is nothing wrong in using letters if you prefer. Similarly, the visibility has been shown using the letters g, m, p and the fog symbol ≡.

After the sea area weather, the next stage is to plot the weather reports from the coastal stations using a standard 'model'. This

Fig. 17 The information from Fig. 16 plotted on to the Metmap. Expected changes in the weather over the sea areas could also be shown but this makes the presentation very cluttered. Positions of fronts and the low have been updated from the synopsis and by the use of other data such as observation by the yachtsman and the content of general forecasts.

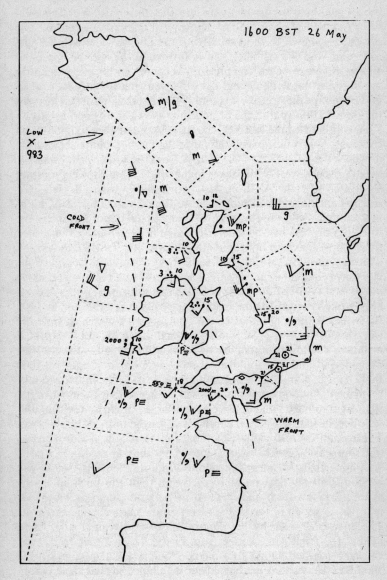

model has the wind arrow flying into the station dot with the wind direction and the wind force shown by feathers as before. The pressure is shown to the top right of the dot with the weather just to the left and the visibility a little further to the left. The pressure change is shown below the pressure. There is nothing significant in the order of the symbols and the other information any more than in the symbols themselves. It is simply that a standard form to present the information is convenient and that shown is as good as any.

Having plotted the weather, the chart is already beginning to look coherent. You will immediately note the flow of winds around the low over the Atlantic and the change from the rather misty weather over the South-west Approaches to the brighter but showery weather associated with the north-westerly winds over western areas.

The final stage of the plotting is to decide where the lows, highs and troughs were at the time of the coastal reports, this is the updating process mentioned earlier. As far as the low is concerned the synopsis gave an approximate position at 1300 BST today and a forecast position for 1300 BST tomorrow. From these two positions it is easy to see how far the forecaster thinks that the low will move in the twenty-four hour period. From this information it is reasonably straightforward to interpolate a position for the low at 1600 BST, that is three hours after the synopsis time by which time the low should have moved about one-eighth of the total expected distance. The low is deepening by eight millibars in the twenty-four hour period and so should be about one millibar deeper by 1600 BST. This one millibar deepening might seem to be an unnecessary refinement but sometimes a low will be deepening by much more than eight millibars and in three or four hours might be several millibars deeper than the actual depth given for the time of the synopsis. The trough over Rockall, Shannon and west Sole is described as 'moving east steadily' which means in the range 15–25 knots. Assuming that this means at about 20 knots then this trough will be about 60 nautical miles further east than the position in the synopsis. Is it a front and, if so, of what kind – cold or warm? Or is it just a trough in the isobars with no air mass change across it? The answer is given by the description of the weather ahead and behind the front; the change from misty, dull weather to clearer, showery weather is typical of a cold front. The Valentia report for 1600 BST shows that the barometer is falling and that there is rain and drizzle with a poor visibility. This is consistent with a cold front approaching Valentia which, itself, is still in the warm air. Had the barometer begun to rise, the rain stopped and the visibility improved then the cold front would have passed through Valentia. If the cold front had still been a long way to the west then it is more likely that the pressure at Valentia would have been falling only slowly and that there would have been drizzle but no rain; there might not even have been any drizzle but the visibility would have been moderate at best and more likely poor.

Is there a warm front? One was not mentioned in the forecast but remembering that the warm front should be leading the cold front might make us wonder why there was rain being reported at Malin Head, Tiree and at Ronaldsway. This cannot be the cold front, which is more likely to be a quite narrow band of rain and is unlikely to have moved from being over Shannon at 1300 BST to be giving rain over the Irish Sea at 1600, unless it is moving very much faster than stated. The poor visibility at Scilly and the Channel Light Vessel as opposed to, say, Royal Sovereign, Varne and Noordhinder suggest that the first two are in a warm sector and the others are ahead of the warm front. Perhaps the warm front is along a line from somewhere near Malin Head, down the Irish Sea and to sea area Wight.

Why did the forecaster not mention the warm front in the synopsis? This may have been partly because he was concerned with the time and number of words available to him in the forecast and, partly because there may not be very much of a wind change associated with this front. The difference across the front might well be only in terms of rain and visibility and, in the interest of brevity, the forecaster is able to describe the weather over large areas both ahead of the warm front and in the warm sector as 'rain or drizzle', until the cold front has passed when 'showers' might be more appropriate. One answer would be to have stayed tuned in to Radio 4 to have heard the general forecast for the whole of the United Kingdom in which the forecaster from the London Weather Centre might have made a mention of a warm front or, perhaps, one front crossing Wales and south-west Scotland and another approaching the west coast of Ireland.

Having decided that the cold front was near to Valentia and that a second (warm) front was over Wales, while a low was positioned somewhere (about 250 miles) to the west of Bailey we can now think about some isobars. When trying to draw isobars remember that the general run of the winds will be along the isobars and that for 1600 BST the general run of the winds will be given by the winds at the beginning of the forecast – those that we have plotted for the sea areas and those reported at the coastal stations. When writing his script the forecaster has been looking not only at his last analysed chart and at the forecast chart for tomorrow but also at his latest reports from ships and from a large number of land stations. The conditions which he describes at the beginning of the shipping forecast are, to a large extent, an extrapolation for a few hours from his most up to date charts. The wind pattern which he describes for the initial stages of the forecast is thus very much a description of the isobaric pattern with only minor adjustments to allow for a small movement of the weather systems and their development.

The isobars over the areas Fastnet, Lundy, Irish Sea and Malin will run in a general direction from south-west to north-east, but how far apart will they be? We know that the wind speed is determined by the isobar spacing and that the forecaster has given winds based upon his latest chart. By using the scale marked Beaufort Force on the map on Fig. 11, p. 59 we can relate isobar spacing to wind speed. For a force 5, for example, place one point of your dividers on the line on the scale by the letter B of Beaufort and the other point of the dividers on the line corresponding to force 5. The distance apart of the dividers is now the spacing of the isobars at two millibar intervals on this scale of map for a force 5 wind. The wind in these areas is expected to be force 5 or 6 and so if we set the dividers to a corresponding spacing and then draw some isobars to fit the pressures at Valentia, Malin Head, Ronaldsway, Tiree and Scilly then the result should be something like that part of the chart in Fig. 18.

In the previous paragraph emphasis was placed upon drawing the isobars in accordance with the 'forecast' winds rather than the 'actual' winds reported from the coastal stations. The reason for this is that at a number of the reporting sites the wind might be affected by topography. Even at Scilly the wind reported can be significantly less than that over the open sea. At Bell Rock the direction of the wind is often affected by the nearby coastline and at Royal Sovereign the sea breeze can make the wind unrepresen-

Fig. 18   Partially drawn isobars and fronts as described in the text.

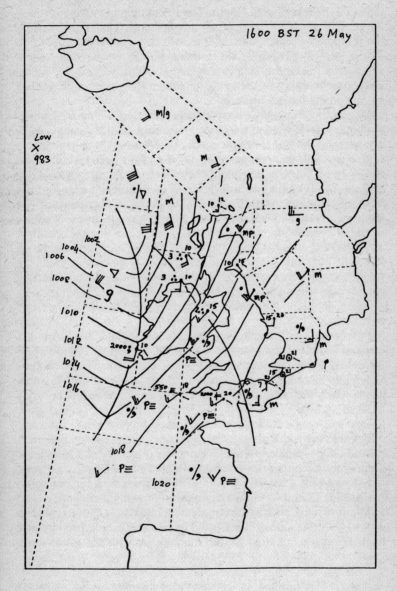

tative of the rest of area Wight. The forecasters are aware of these effects and allow for them when drawing their charts and writing the shipping forecast. Sometimes the winds from these stations will fit the general pattern of the isobars, but be prepared for some anomalies. The coastal station winds may, of course, give the yachtsman some warning of the deviations from the open sea wind that he can expect to find inshore.

On meeting fronts the isobars change direction in a manner which implies a veer of wind at the front; this means that the isobars form a V-shape at the front with the point of the V towards high pressure. In this case this is in accordance with the winds over Hebrides and Bailey being south-east while those over Irish Sea are south-west, and the winds coming in over Shannon and Rockall are from the west or north-west. At the front itself the spacing of the isobars across the front is related to the speed of the front. The scales on the Metmap in Fig. 11, p. 59 show these relationships, again for two millibar intervals, one scale for warm fronts and one for occlusions and cold fronts. In this particular case the cold front is moving at about 20 knots and this gives some idea of the spacing of the isobars as they cut this cold front. The speed of a front as given by the isobar spacing at any point along the front is the speed perpendicular to the front at that point.

For the sake of completeness the isobars around the centre can be filled in by knowing the pressure at the centre and interpolating from there to the reported values at Tiree, Malin Head and Valentia. When this has been done the spacing of the isobars should indicate winds which are not too different from those forecast for the areas Rockall and Bailey. Similarly, over the North Sea the isobars must follow the winds and have a corresponding spacing, the only difference here is that there are no pressure centres to help the interpolation. The same applies over the areas of Biscay and Finisterre. Sometimes, of course, the forecast will mention other pressure centres in the synopsis: a low over Spain or the Azores high for example.

Figs. 17, 18 and 19 show the stages in drawing this chart.

The drawing of such a chart from one forecast in isolation is always difficult whether one is experienced or not and even a professional meteorologist would find the job much easier with the aid of some

Fig. 19 The finalised version with the plotted data omitted for clarity.

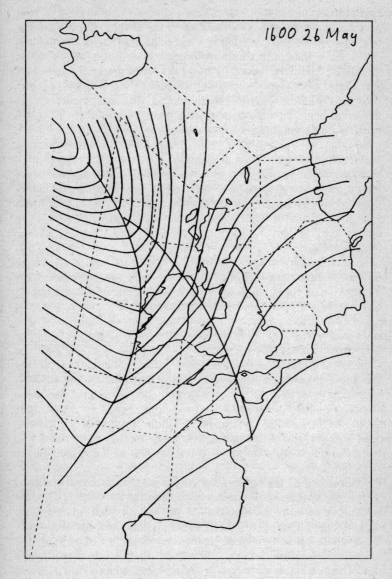

1600 26 May

background information. In practice you would always have your previous chart to give some idea as to the shapes of lows and the positions of fronts. Particularly important are the fronts because, as we have seen, the shipping forecast does not always include mention of them. However, by using the scales marked 'warm front speed' and 'cold front or occlusion speed' at the top of the Metmap the speeds of fronts can be estimated and 'forecasts' made of their movements in the period from one chart to the next. You will probably find that fronts move at varying speeds, faster near the centre of the low pressure than further out towards the higher pressures. Assuming, as a first guess, that the front moves at a constant speed for the four to seven hour period from one chart to the next then the expected position of the front can be lightly drawn on the Metmap. This can then be compared with any information given to you in the synopsis about the front, with any of the coastal reports near the front and with the weather expected over the sea areas.

Of course there always has to be a first chart and this is where the newspapers come in useful, or at least those which publish either a forecast map for the day of issue or an actual chart for noon yesterday. Either will prove useful in helping to get the shapes of the lows and highs as well as the types of fronts. A quick look – and that is all you will get – at the television map can also help if you have a retentive memory. An alternative is to visit one of the Weather Centres, in London or Southampton for example, and ask to see the latest chart or even obtain copies of both actual and forecast charts. Also there is no reason why you should not take down the shipping forecast at home and spend some time drawing a chart or two in the comfort of your home before leaving for your yacht. This will not only help to give you some of the background you will need but it will also serve to remind you of the format of the shipping forecast.

The method of taking the forecast down from the radio and then drawing a chart is one which can be practised at any time by listening to the forecasts on 1500 metres or, if you are not at home, by getting some obliging member of your family to record the forecast for you. Your chart can be checked against the forecasts shown on the television or against the charts published in the press on the following day. The next stage, of course, is to decide just when a cold or warm front will pass through wherever you are and then check with what actually happens. By building up experience in this way the keen yachtsman can begin to recognise some of the types of weather system that the forecaster finds most difficult. This practical approach is by far the best way to learn about the weather and one which will teach you much more efficiently than reading textbooks alone. If you get really good then you may begin to be regarded as a *savant* by your friends and acquaintances as regards your knowledge about the weather and the necessity to take an umbrella to work. Be careful though! Unfortunately the weather does not always do what is expected, and both the forecast and you will err from time to time.

Some Colleges of Further Education run courses on the RYA Yachtmaster's Certificate which includes a section on meteorology. Sailing clubs also run such courses and either may run courses solely on the meteorological part of the syllabus. The RYA publishes a teaching aid, *The Yachtsman's Weather Map*, which has a cassette of recorded forecasts with some brief guidance on how these should be analysed as well as showing the final stage of each chart and the intermediate phases of some of them. The cassette also demonstrates some of the other forecasts which are available and shows how these can be used as well. Like much of sailing there is a limit to what can be learnt by reading, the real learning stage is in doing it yourself. At first even simple navigation and pilotage seem difficult while astro-navigation seems impossible. And so it is with the weather forecast until you really have a go for yourself. It is only then that you discover what can be achieved and what the real difficulties are.

Using the chart

We can now return to the questions posed earlier relating to the forecast, from which the chart in Fig. 19, p. 119, has been drawn. First, the visibility in sea area Wight will improve when the cold front gets down to that part of the Channel. But when will that be? As drawn the cold front is some 400 miles away from the western part of Wight and, at 20 knots, the cold front is going to take some twenty hours to reach the area. If the cold front was moving at the top end of the speed range implied by 'steadily', that is 25 knots, then the time could be as short as sixteen hours, while at the lower end of the speed range the time could be as much as twenty-seven hours. Similarly, the veer of wind in sea area Plymouth will also occur with the cold front and at a distance of 300 miles from the cold front to the middle of the area the time of the veer can now be placed as at about fifteen hours from the time for which the chart was drawn or in the range twelve–twenty hours. These time ranges may seem rather rough and ready but they do at least give some data upon which to plan, and they can be updated when you hear the next forecast.

In this particular forecast the trough, or front as we were able to deduce, was given a position and a speed. However, if these data had not been given you might, nevertheless, have decided that there was a front and been able to fix, within limits, its position. This might have been possible either from a previous forecast, from information given in another forecast or from the coastal station reports. The speed of this front for the next few hours could then be estimated from the isobar spacing across the front. Obviously the accuracy with which you make this estimation will not be very high but will be better than nothing and may be improved in the light of information gained from other sources.

After the cold front has passed across sea area Fastnet the pressure would normally be expected to rise; after all the low is moving towards north Faeroes and not towards Fastnet. If the pressure began to fall after the initial rise on the passage of the fronts then this would almost certainly be for one of two reasons. Either the low could be coming further south than the forecaster thought was possible, perhaps with troughs swinging round in the air behind the cold front, or there might be a wave forming on the cold front. In the first case all that would be likely to happen would be that the wind direction would remain more from a general westerly rather than from a north-westerly point, but backing and veering as the troughs passed, giving some heavy showers or longer spells of rain. These might be unpleasant but not dangerous, inconvenient but not hazardous. In the second case a wave on the cold front can be a rapid, vigorous development of great potential danger especially if it develops into a low in its own right. Some waves do not behave in this way but the occasional one does and it is not unknown for severe gales to occur at very short notice. Pressure falls in such an area, in the vicinity of a cold front, should always be treated with the greatest caution. The best advice is to be on your guard, listen to the next shipping forecast, keep a listening watch for gale warnings (this is where the PO coastal radio stations are so valuable) and watch the barometer for further falls, especially if these increase. The real warning will be a marked backing of the wind and some increase in its strength, and the former, rather than the latter, is the real give-away.

One of the important uses of the analysed chart is in obtaining some idea of what the forecaster means by his very general descriptions of the winds. In this forecast the wind coming in behind the cold front was described as 'west to north-west 5 or 6'. This covers anything between south-west and north in the speed range of force 5 or 6. The user might want to know just how far to the south of west the wind will be in these western areas and whether the wind is going to vary between the directions mentioned or whether there is going to be any trend from one to the other. If there is going to be a trend then why did the forecaster not say 'west becoming north-west'? To fit all the data, with the low still to the west of sea area Bailey, the isobars could not easily be drawn to show very much of a north-westerly. Indeed, a wind from that direction would probably only occur as this low was passing to the north of Britain. You will note that behind the cold front over the Irish sea the wind was forecast to be north-west – no mention of westerly here! The answers to these questions then are that the wind over the western sea areas is unlikely to be very far to the south of west after the passage of the front, or if it is then not for long. But, as the low carries on north-eastwards so the winds behind the cold front will become more of a true north-westerly and, therefore, there appears to be a trend. Perhaps the forecaster did not mention the trend because he is concerned that the developments might not be quite as clear cut as he would like and he is hedging his bets. He may be trying to say no more than he really knows!

From experience in trying to draw charts from the forecast you should be able to acquire some idea about what shapes are possible for the isobars and fronts. This will enable you to get a clearer idea about what the forecaster means when he describes the wind over a group of areas by means of a range of speed and directions.

The example of this shipping forecast shows how it is possible to derive a great deal more from the forecast than the forecaster can possibly have time to say. The deductions made about the timing of the improvement in the visibility or the veer of the wind could, for example, be used in a strategic sense to decide whether to delay leaving harbour or not, or, in a tactical sense when on passage, to decide whether or not to tack. One specific example of the first type of decision could be if you were in Cherbourg and wishing to cross to the Solent. You might not be happy about crossing the shipping lanes in a poor visibility but would know that a delay of fifteen hours or so would allow the visibility to improve. There might be good reasons, of course, why you could not delay your departure. A second example might be if you were at anchor in the Fowey River and intending to make a passage across the Bay of Biscay. To start right away would mean that you would be faced by headwinds and a poor visibility. Knowing that a veer of wind was expected with an improvement in the visibility might suggest that there would be no penalty in delaying your departure.

The tactical sense of the use of forecasts might be illustrated if you imagine yourself having left the South Coast and crossing the Channel, intending to round Ushant and proceed across the Bay of Biscay. On the starboard tack you would, with the wind so far round to south, clearly not be going to lay Ushant and, therefore, you would have to decide when to tack on to port. If you were some distance from Ushant then the front and the attendant wind veer might reach you sufficiently early to allow you to head up to windward and so avoid having to tack. By calculating the course that you would make if the wind veered the least amount, say to west or west by south, then you can decide how long it is worth holding on to the starboard tack knowing that the veer would definitely take you around Ushant. You might, of course, decide to tack somewhat earlier so that when the veer came you could make best use of it and, if it veered enough, perhaps even use the spinnaker.

From the analysis of this forecast some indication has been given of the types of question which can be asked – and answered. You should not, of course, rely on one forecast alone. You should be monitoring the weather continuously, not only by observing what is happening to the wind, barometer and weather but also by monitoring the forecasts and listening for changes in emphasis from one to the next. This is much the same as making navigational checks when you might be comparing your dead reckoning with astro-fixes or bearings from a radio compass. Are the fixes consistent with your course heading and speed? Does the radio bearing confirm or make you query the DR? Should the barometer be falling? Should the wind still be south-west? Why has it not yet veered? The philosophy is the same: check, check and check again!

Differences between observation and expectation can give early warning of the forecast going wrong. Most forecasts have errors of some form, many of which will not worry you, but the occasional one will and then all the trouble is worthwhile. A falling pressure rather than a rising one, especially near a slow moving cold front is often a danger signal as is a backing wind instead of either a veer or a near constant wind direction. An accurate barometer is useful when checking the forecast. First, it can give you another pressure to help you draw your chart. Secondly, if you can hazard a guess as to how low the pressure is going to be over the next few hours, by assuming that the whole pattern of isobars is going to move with little change in shape, with the low and high centres, then, if your barometer has been calibrated recently, you should be able to see whether the pressure is much lower than expected. If the weather is obviously not proceeding according to the forecast then try to see for yourself what is happening and whether it could be hazardous, keep a listening watch out for gale warnings and make sure that you do not miss the next shipping forecast. In this way you will stand a chance of keeping one step ahead of the weather or, sometimes, just that half step ahead that can make all the difference. When at sea you will not often be able to avoid bad weather but at least you will be able to make everything secure, check all your safety equipment, make sure that you have enough sea room and have some good navigational fixes.

By interpreting forecasts rather than by using the forecast at its literal face value or, worse, not using it at all, you can be that much safer at sea. You may also, on some occasions, be able to proceed in the knowledge that the gale warning issued for your sea area is unlikely to affect the part where you are. In short you can make much more intelligent and more efficient use of the forecasts and so get far more and safer enjoyment out of your sailing.

A final reminder about the shipping forecast and gale warnings; winds are reported and forecast using 'true' directions and not relative to magnetic north. This can lead to winds mentioned in the forecast being, apparently, more in error than they really are because the yachtsman always has in front of him a compass reading directions from magnetic north. The wind thus might appear to be veered by about 10 degrees upon the direction in the forecast or deduced from the Metmap. The yachtsman has to make a conscious effort to change from the one system to the other.

7

Other Sources of Information

Although the shipping forecast is the main source of weather information to the sailor there are other forecasts which can be used to supplement and complement it. Some of these other forecasts are for the general public and some are for specialist use. These other forecasts will be itemised and their use to the sailor discussed.

The Inshore Waters forecast

This forecast is broadcast once a day on Radio 4 immediately after the late night shipping forecast and is intended for inshore shipping and sailing. The presentation is similar to the shipping forecast in that it includes a synopsis, forecasts for areas around the coast of the United Kingdom to a distance of twelve miles out to sea and, finally, reports from some stations around the coast. There are two versions of this forecast, one for broadcast by the English and Welsh Radio 4 programmes on 1500 metres and by BBC Radio Scotland. The second version is broadcast by the BBC Radio 4 Ulster programme. The coastal areas covered and the coastal reports given in each forecast are appropriate to the individual programmes. The coastal reports included in the Inshore Waters forecast are from stations which are not broadcast in the shipping forecast.

The use of the Inshore Waters forecast is thus twofold; first it can give information relating slightly more specifically to the waters near the coast where the majority of yachtsmen are likely to be and, secondly, it can give extra actual weather information to help draw isobars and fronts from the 0015 Shipping forecast. The principal disadvantage of this forecast is that it is only broadcast once a day. Use it, therefore, as an adjunct to the 0015 Shipping forecast and do not assume twelve or eighteen hours later that the forecast you heard the previous night is still valid. However, the next forecast mentioned will give useful updated information.

Early morning forecast on Radio 3

This forecast is broadcast on the medium wave channel of Radio 3 when the station first opens, 0655 on weekdays and 0755 at weekends. It is in several parts and includes a synopsis ot the weather pattern as well as an Inshore Waters section. The latter is similar to the forecast part of the Inshore Waters forecast but does not include the coastal reports. Since the change in radio frequencies in 1978 Radio 3 has only been broadcast on one medium wave frequency and so this forecast might be difficult to receive in some areas.

General forecasts on Radio 4

These forecasts are primarily for the use of the general, land based, public and are intended to describe the weather expected to occur over the whole of the United Kingdom. Because of the variability of the weather for the very many reasons which have been mentioned in this book the treatment in this forecast is, necessarily, of a very general nature. The emphasis is naturally on the major centres of population or areas of particular interest. The main use of these forecasts to the yachtsman is to give a little more information than can be included in the shipping forecast. In particular there may be some indication regarding the uncertainties of the weather developments as well as extra information on the present weather. There may be reference to warm or cold fronts and their positions or some mention of rain reaching parts of the mainland or a clearance following rain, for example. These can help in the positioning of fronts.

A particularly valuable feature of this forecast is the outlook for the following forty-eight hours. This can be of great use to the yachtsman when he is making decisions about his best course of action. For example, he may wish to make a passage within the next day or two and the wind for the first twenty-four hours is forecast to be stronger than he likes although not unmanageable. The outlook forecast may give some idea about the possibility of the weather becoming more or less disturbed, in the first case he might decide to leave right away rather than risk facing even stronger winds later. In the second case it could well pay to wait in harbour for a further day. However remember that, in general, the further ahead the forecast the more inaccurate it is likely to be and so the outlook forecast has a greater likelihood of error than the forecast for the first twenty-four hours. For this reason the outlook is usually phrased in even more general terms than the main part of the forecast.

Local radio stations

The service presented by these stations varies greatly. Some stations, Radios Solent, Cleveland and Newcastle, for example, give forecasts which are tailor-made for mariners while other stations only give forecasts for land areas. Yachtsmen are advised to check with their local radio stations as to what services are available and, where there are none, or where the service is inadequate, there may be a case for local yacht clubs and other marine interests to request the local station controller for more time on the air.

Automatic telephone weather service

These forecasts are prepared by the Meteorological Office and are among the PO recorded services available to all telephone subscribers. Usually these forecasts are designed for land-based activities such as picnics and hanging out the washing; as such they reduce the work load of the staff at weather centres where many telephone queries are handled directly by forecasters or their assistants. In coastal areas the ATWS forecasts contain some information of value to sailors who intend to stay close inshore; for example, they can alert the yachtsman to the possibility of such hazards as squalls and fog. One drawback to these forecasts is the delay in updating them when the forecast requires amendment. It has been emphasised elsewhere in this book that forecasters keep the weather under continuous review and that they are constantly considering the need to amend forecasts in the light of the latest information. An ATWS forecast issued some four or five hours ago may well be due for amendment or for a routine updating and extension forwards in time. The user should listen to the time of issue of these forecasts and if it is more than three or four hours old he should use his own judgement to decide whether the forecast is likely to offer good advice or not. Compare the ATWS forecast with the weather that is occurring at the time the call is made; this might give an indication as to whether the rest of the forecast is likely to be seriously in error or not. Probably the best use of the ATWS is to help the yachtsman in deciding whether or not to make use of the next facility available to him – the personal telephone call to the forecaster.

Personal telephone briefing

The Meteorological Office maintains a number of Weather Centres whose particular functions include the provision of forecast services to the general public and there are other forecast offices on RAF and civil airfields which may, other commitments permitting, be able to answer telephone queries from the public. Telephone numbers of these can be found in PO telephone directories, the RYA booklet *G5* and in two Meteorological Office leaflets – *Weather Advice to the Community* and *Weather Bulletins, Gale Warnings and Services for Shipping*. (Leaflets numbers 1 and 3.) When at sea a vessel with the appropriate radio equipment can speak direct to the forecaster at Bracknell using a medium or high frequency R/T link or, alternatively, requests can be relayed via the PO coastal radio stations. Whenever use is made of these services it is as well to remember that the forecasters deal with very many diverse enquiries during the course of the day in addition to handling routine forecasts to a tight time schedule. In order to get the best from the service you are advised to do the following:

(i)   Say, reasonably precisely, what you are doing, or want to do, and where.

(ii) Make sure that you already know something about the current weather pattern.

(iii) If you are some distance away from the forecast office be prepared to tell the forecaster what the weather is doing at the time of your query.

(iv) Do not hesitate to question the forecaster, particularly regarding his degree of confidence or otherwise in the forecast, but be brief and to the point, he may have other people waiting.

Above all – listen carefully!

Some of this advice may surprise the reader. Taking the various points in turn, first, if you tell the forecaster what you are doing then he will be able to concentrate upon the weather elements which will concern you and he will have some idea of the limitations under which you wish to work. For sailing he will know that winds of force 6 and upwards will be of concern to you as will visibilities below two miles or so. Rain, which might be of paramount importance to a person organising a garden fête will only concern the sailor if it is heavy enough to impair the visibility or if there are heavy showers with strong gusts.

Secondly, if you know a little about the current weather pattern then you are more likely to be able to ask questions intelligently and so benefit the more from the briefing.

Thirdly, the forecaster is always working with data that cannot be much less than an hour or so old, and may well be older, from a network of observations which is never as dense as he would really like. To get some up to date information from an area where he may have no reports can be of help to the forecaster and may, additionally, save time – he may not have to discuss the rain which may reach you within the next hour or so if he learns that it has already arrived!

Lastly, every forecast is an expression of probabilities upon which the user bases decisions. As the user your decision-making could be influenced by the certainty or otherwise of the forecast.

Shipping forecasts from other countries

Details of these are to be found in the *Admiralty List of Radio Signals, Vol 3* and in the RYA booklet, *G5*. The format is much the same as the British forecasts and some are in English. The two publications give the areas used and *G5* gives translations of the technical terms appearing in those forecasts that are broadcast in the language of the originating country.

PO radio stations

In addition to broadcasting gale warnings, as described in Chapter 5, these stations also broadcast twice a day a shortened version of the shipping forecast. This includes gale warning summaries, the general synopsis and area forecasts for the areas near to the station – the same areas for which gale warnings are broadcast. Depending upon the time of year, that is whether the United Kingdom is using GMT or BST, these forecasts are broadcast between two and three hours *after* those shipping forecasts which are broadcast on Radio 4 at 0625 and 1750. The forecaster thus has the opportunity to amend the original forecast before it is broadcast by the PO station. Schedules are to be found in the *Admiralty List of Signals*, the RYA booklet and the Meteorological Office leaflet no. 3. A drawback to this otherwise very useful service is the fact that the forecast is only broadcast twice a day. Another disadvantage is the limited amount of information which is broadcast.

Press forecasts

Because of the delay between writing them and the time that they are read these forecasts are often of minimal use to the yachtsman. In those papers which are printed in London and then despatched to the west country, for example, the forecast may have been written at noon on the previous day. The forecast in the national dailies is issued by the Meteorological Office, at the very latest, at 2300 the previous evening and, in most editions, by 1900. Some papers publish a weather chart which may be either a forecast chart for noon of the day of issue of the paper or the actual chart for noon on the previous day. In either case the chart is very useful background information for the yachtsman about to set sail and can be a valuable aid to the understanding of the shipping forecasts.

Television forecasts

Similar remarks apply to the television forecasts as to those in the newspapers but for slightly different reasons. The brief form of the forecast on the television makes it necessary for the forecaster to concentrate upon the major centres of the population and any really severe or notable weather but the presentations usually include a sight of a chart of isobars and fronts, sometimes with a satellite cloud picture as well. This can again be useful background information to the yachtsman.

W/T Transmissions

If you are able to read Morse code then you might be able to make use of the forecasts broadcast from Portishead by the PO or by the Royal Navy on the Whitehall frequencies. Details of these transmissions and the contents of the broadcasts are to be found in the *Admiralty List of Radio Signals, Vol. 3.* For many the speed of transmission will be too great for their use to be a practical possibility and they do demand some knowledge of the international meteorological codes. However, these broadcasts give a wealth of information on actual and forecast patterns of isobars and fronts as well as actual weather reports from ships and land stations near to the seaboard. Anybody contemplating trans-oceanic sailing should, if possible, learn morse sufficiently well to make use of these forecasts which are much used by professional mariners.

Facsimile broadcasts

The Meteorological Office, in common with the national weather services of other countries, broadcasts weather charts on radio facsimile. These charts are for the use of commercial shipping and the Royal Navy and their reception needs specialised equipment. Such equipment is now becoming available suitably miniaturised for its use to be possible on board small boats and, with continuing technological advances, further miniaturisation is, no doubt, possible. Details of schedules, frequencies and broadcast content can be found in the Admiralty *List of Radio Signals*.

Volmet

Anybody with a radio capable of receiving VHF transmissions intended for aircraft can receive Volmet. This is a continuous broadcast giving weather reports for a number of airfields over the United Kingdom and western Europe. The airfield name is followed by the wind, in degrees and knots, visibility, weather and cloud – if these are of significance to the aviator. If these are above certain limits the report simply says CAVOK (pronounced cav-okay) meaning cloud and visibility OK! The sea level pressure is given in millibars and is preceded by the code letters QNH. The pressure and the wind reports if recorded for the time of the shipping forecast coastal reports, could assist in the drawing of isobars from the shipping forecast.

PO Viewdata (Prestel)

This service introduced by the PO will make available to all telephone subscribers (with suitably modified television sets) a wide range of information including weather forecasts and, in particular, the shipping forecast. In time many yacht clubs will, no doubt, have the necessary equipment installed, as will members of the public. With Prestel the subscriber can present upon his television screen the shipping forecast to be read at his leisure as well as actual weather information. At present the graphics available on Prestel are relatively crude and isobars cannot be shown. This service is likely to develop greatly over the next decade.

Weather lore

Another source of additional weather information is contained in so-called weather lore. The word 'lore' gives the impression that the sayings which come under this heading are little better than old wives' tales, and in some instances this is so. Many examples of weather lore are, however, based upon shrewd and careful observation. Under the first category must come such sayings as relate to St Swithin's Day or the severity of a winter being dependent upon the number of holly berries. In the second category some sayings such as the 'Red sky at night, sailor's delight, Red sky in the morning, sailor's warning' couplet have more than a grain of truth. A red sky at night implies that the sun's rays coming over the western horizon are unimpeded by cloud to the west and so can illuminate the underside of clouds in the vicinity of the observer. As the weather usually comes from the west it is likely that the next twelve hours or so at least will be fine. Red sky in the morning is probably less reliable in that it implies that there is fine weather to the east and, therefore, because of the characteristic dimensions of weather systems, a good chance of unsettled weather not being too far away to the west.

'Rain before seven, fine before eleven' is simply an expression of the fact that rain belts are often of such a size that they pass over in a time which averages less than four hours (but which may be considerably more). The times seven and eleven merely rhyme and the same phrase could be used with any two clock times three to five hours apart, but with less poetic appeal.

A backing wind is often a sign of low pressure approaching and hence the couplet 'When the wind shifts against the sun, trust it not for back it will run'. A strong pressure rise behind a cold front may well indicate that the wind in the cold air will be strong and this is expressed in the saying 'Strong rise after low foretells a stronger blow' or, because the cold air will be showery and there would probably also be strong gusts, this rhyme might also be 'When rise commences after low, squalls expect and then clear blow'. The fact that sea breezes occur in fine settled weather and that the sea breeze veers during the day is noticed in 'When the wind follows the sun, Fine weather ne'er be done'.

Slow moving weather will take a long time to pass over and fast moving weather will pass quickly and hence the rhyme 'Long foretold, long last, short notice, soon past'. Fine weather is often indicated by the so-called 'mackerel sky', this is a cloud type known technically as cirro-cumulus or alto-cumulus. If the little globules of cloud are high up and changing little then this is a good sign recognised by 'A mackerel sky, let all your kites fly'. If a change is on its way then the cloud may change appearance and 'Mares' tails' will spread across, these are still, high, streaks of wispy cloud – the first signs of warm front cirrus. The rhyme 'Mackerel sky and mares' tails, Make tall ships carry small sails' describes this effect.

Some sayings refer to 'halo' size, the larger the halo the more likelihood of rain, the smaller the halo the more chance of fine weather. The reason here is that the large haloes occur with a uniform sheet of cloud high up in the sky and this, typically, is the cirro-stratus ahead of a warm front (see Plate 4). In fact the halo is often associated with a fairly slow but steadily moving warm front. The 'small' haloes are really the phenomenon known as coronas caused by small water droplets and which usually occur with the 'mackerel' sky type of cloud, the cirro-cumulus or the alto-cumulus. Coronas are thus indicative of fine weather.

Most of the weather lore mentioned and, indeed, most of the sayings that have any validity relate to weather which is fairly imminent. A few of the slightly longer period sayings have some scientific basis and some of these are those which refer to the persistence of easterly winds. This is a high pressure phenomenon in many, but not all, cases and one which gives rise to sayings of the form 'If the wind be north-east, three days without rain, eight days will pass before south again'.

Contrary to popular conception there is no scientific evidence that the weather is dependent upon the moon or on the changes in the phase of the moon. There is, in fact, a rhyme which states this – 'The moon and the weather may change together, but a change in the moon does not change the weather'.

The fact that some of the sayings which come under the general title of weather lore can be found in the writings of Aristotle suggests that they have stood the test of time and do have some predictive value. They are, in fact, just another source of information of which the sailor should be aware. They point to the need to observe the weather and to relate what is observed to what follows. A halo observed when no rain was forecast or a backing wind could well be the first sign that the shipping forecast might be in error. Just because there are sophisticated techniques for forecasting the weather the user of forecasts should not eschew the old rules, they still have some limited but nevertheless valuable use.

The last couplet 'The lyf so short, the arte so long to lerne' will simply serve to remind the reader that meteorological knowledge will not be gained overnight by a quick read of this or any other book; there is no instant road to success.

Weather lore represents the distillation of the experience of centuries; another form of lore of more recent origin is contained in some of the popular ideas of why or in what weather patterns forecasts go astray. Members of the public can sometimes be heard to say that forecasts are often 'slow', that is they get the pattern of weather correct but the timing wrong. There is some truth in this assertion although there are cases when the forecast is too 'fast'. Errors in timing of the major weather systems, lows, highs and fronts occur for two main reasons. First the forecaster might not know sufficiently precisely the initial position of these systems at the time when he is preparing the forecast – an initial error in position of 60 nautical miles in a system moving at a typical speed of 20 knots can clearly give an error of three hours in the time of arrival of a front with its attendant wind shifts. An error of 60 miles in the position of a front over the Atlantic Ocean, even with the use of satellite pictures, is not unusual if only because of the ill-defined nature of some fronts. There is, therefore, a random error likely in the timing of weather systems because of the uncertainty factor in the forecasting process.

There are, however, two types of weather pattern when timing may be more suspect than at others. One of these is when there has been a high pressure system over the United Kingdom and Europe that is blocking the eastwards progression of the frontal lows which are approaching across the Atlantic. There are always difficulties in trying to decide how far into the high any individual system will encroach. The real problem to the forecaster is, however, when the high is showing signs of collapsing. These patterns always look so stable and yet the forecaster can see that the collapse is about to take place, it might be with the next frontal system or the one after. Often, but not sufficiently often for it to be a rule, the high gives way more quickly than expected and the forecaster has been over-cautious in his estimate of how quickly the fronts will cross the United Kingdom. The converse to this situation is when a high develops and starts to block the flow. Again the forecaster can often see what is going to happen but sometimes the high develops a little quicker than expected and the forecaster has been over-hasty in his prediction of the movement of the frontal system across Britain.

Like all forms of meteorological knowledge experience and observation are of prime importance and the sailor can only be encouraged to become aware of those patterns which are more likely to be associated with errors in the forecast.

The main lesson to be learnt from this book and, perhaps, particularly from this last chapter is that there is no single tool for the sailor to use when he is trying to decide what the weather may or may not do. The various weather forecast services should not be used as 'stand alone' information but must be used in conjunction with observations of the weather, wind and pressure. Conversely, rules of thumb, such as weather lore and other so-called rules which attempt to describe what the weather will do, are doomed to failure if used in isolation. Single observer forecasting, as it is called, has strictly defined limitations by virtue of the amount of sky which can be seen from a point and the distance ahead of a weather system at which there may be observable effects such as pressure falls and increasing cloud.

The benefit which will accrue to a yachtsman from using weather information will be dependent upon the effort which he is willing to devote to understanding the problem and, subsequently, the attention which he is willing to give to the weather while sailing.

Finally, the yachtsman and, indeed, any other person using a weather forecast must always bear in mind that it is no more than a statement of opinion based on the evidence available to the forecaster at the time. It represents the best that can be done with the data and the techniques available. There is not and cannot be any guarantee of accuracy and there is no assurance that new data will not lead to a different forecast. The nearest, and most often quoted, parallel is the medical prognosis which, like its meteorological counterpart, depends initially upon an incomplete set of data. In both cases the prognosis is, in the final analysis, a subjective opinion based upon the most up to date scientific knowledge of a still incompletely understood system.

The doctor has one advantage over the meteorologist in that he can on many occasions influence the behaviour of his subject. The meteor, logist has no influence over the weather!

Appendix 1

The Organisation Behind the Weather Forecast

The meteorological services of the world are members of the World Meteorological Organisation (WMO), a United Nations agency. Through WMO, standards of weather reporting and coding are agreed internationally so that a weather report from anywhere in the world and written as a set of five figure groups is immediately comprehensible to any meteorologist whatever his nationality and whatever the country of origin of the message. Barometers and thermometers are read, descriptions of weather and cloud are made, winds are estimated or measured and visibilities are estimated at agreed times every day. These data, in the five figure number codes, are then transmitted from individual reporting stations to regional or national collecting centres and thence to a high speed global telecommunications circuit. The speed of this process is such that the weather over the whole of the British Isles can be known to the forecaster at Bracknell (The United Kingdom Meteorological Office Headquarters) within half an hour of the weather being observed. He can have a chart with weather symbols plotted for much of Europe and the eastern Atlantic in just over the hour and, two and a half hours after the observation time, the Bracknell computer can start to work on weather data from all around the northern hemisphere.

The computer produces guidance in various forms from which the forecaster has to prepare both written forecasts and charts describing the weather over the next few days. The state of the science is not yet sufficiently advanced for all the computer products to be used with no interpretation or amendment by suitably experienced and trained forecasters although some specialised aviation forecasts are passed directly to the user.

The forecast weather charts, some of those produced by the computer and some of those amended by the forecaster, are broadcast in pictorial form over the same worldwide circuits which are used for collecting the raw data. Any forecaster anywhere in the world can receive the charts from Bracknell if he needs to. The forecaster at Bracknell receives charts from the United States, Germany, Russia, Japan and Canada on a routine basis and can receive other charts from other countries on request. There is complete international exchange of both raw and processed data; the weather does not recognise national boundaries and meteorology is one area where there is harmony between nations.

In Britain the Meteorological Office is responsible for providing a weather forecast service to the Royal Air Force which is why, nowadays, the Office is a part of the Ministry of Defence. Originally, because of the mainly marine application of weather forecasts at that time, the Office was attached to the Board of Trade which, in its present guise as part of the Department of Trade and Industry, still has a say in the form of presentation of the shipping forecast, as we saw in Chapter 5.

In addition to its roles relating to aviation and marine forecasting the Meteorological Office supplies a free forecast service for general public use as well as forecasts for a number of specialist users such as industry and including the North Sea oil industry, the public utilities and agriculture. There are also consultancy services available to designers and planners. All these services, many of which are revenue earning, are backed up by an operationally orientated research programme. From this it will be seen that the public service work, such as the forecasts supplied through the media of the radio, television and the press, represents a relatively small but, nevertheless, important part of the total work load of the Meteorological Office.

Appendix 2

Some Meteorological Terms Explained

This appendix is not intended to duplicate the glossary, rather it is intended to amplify some of the terms used in the book most of which were defined when first introduced.

Throughout the text the word 'low' has been used for a low pressure area. The word 'depression' may be used as being synonymous with low. The circulation of winds around a low is known as 'cyclonic' and this term is used in the shipping forecast to describe the winds around the centre of a low, but the word 'cyclone' is usually reserved for the tropical depressions which are not sufficiently vigorous to be known as typhoons or hurricanes. The 'pecking order' of tropical disturbances is hurricanes or typhoons followed by cyclones and then tropical depressions.

The converse of low or depression is the 'high', as used throughout this book, or the 'anticyclone'. The word anticyclone can be used for high pressure areas regardless of latitude and may be heard in general forecasts but not normally, because of its length, in the shipping forecast. The circulation of winds around a high is called 'anticyclonic'.

Fronts, or frontal zones as they are rather than sharp demarcation lines, are well marked areas separating air masses of contrasting origins. Because of the ascent of air taking place in association with the air mass boundary there is a measure of inflow or convergence of air at ground or sea level into the front. The term 'inter-tropical front' used to be used to describe the zone where air masses from the northern and southern hemispheres meet, as described in Chapter 1. Because there is no essential difference in the two air masses in terms of their temperature and humidity structure, the word 'front' is nowadays considered to be inappropriate and this area is known as the 'Inter-Tropical Convergence Zone' (ITCZ).

The word 'forecast' was first used in the meteorological sense by Admiral Fitzroy who was in charge of the Meteorological Office when the first weather forecast was issued in 1861. The word 'report' is sometimes used by members of the public to mean forecast. To the meteorologist the word 'report' is synonymous with actual weather which is occurring or which has occurred.

'Veering' and 'backing' describe the change in wind direction when the wind changes in the same sense as the hands of the clock in the first instance and against in the second. The terms are used in the same sense in either hemisphere. This means that winds will back ahead of a front in the northern hemisphere and veer behind the front. In the southern hemisphere the wind will veer ahead of a front and back after it. Similarly, gusts, which normally veer in the northern hemisphere, will back in the southern hemisphere.

Appendix 3

List of Books and Other Publications

1   *Weather Advice to the Community* (Met O Leaflet No. 1)
2   *Weather Bulletins, Gale Warnings and Services for the Shipping and Fishing Industries* (Met O Leaflet No. 3)
3   *Sail Weather Wise* (A Meteorological Office leaflet)
The above are all available from the Meteorological Office, Met O, 7, London Road, Bracknell, Berkshire, RG 12 2SZ and are supplied free of charge. Some are also available at Weather Centres.
4   *The Weather Plotting Chart* (Met 07, Form 2), for writing down forecasts in their geographical context. This is available from Weather Centres at a modest charge and can be obtained by post from Southampton Weather Centre, 160 High Street, Southampton, SO1 0BT.
5   *Meteorology for Mariners*, available from HMSO. A comprehensive textbook on meteorology written for professional seamen.
6   *The Marine Observer's Handbook*, also available from HMSO. A guide to weather observing at sea.
7   *Weather Forecasts*, RYA booklet *G5*, David Houghton. A summary of weather forecasts and terms used.
8   *The Yachtsman's Weather Map*, Frank Singleton and Keith Best, available from the RYA, Victoria Way, Woking, Surrey GU21 1EQ. This is a teaching aid on the use of the shipping forecast and includes a cassette tape of forecasts.
9   *Metmaps*, for both recording shipping forecasts and drawing weather maps. Available from the RYA and the Royal

Meteorological Society, James Glaisher House, Grenville Place, Bracknell, Berkshire, RG12 1BX.

10 *Your Own Weather Map*, C. E. Wallington, available from the Royal Meteorological Society. This is a collection of articles on the analysis of the *Metmaps*, perhaps rather too detailed for most yachtsmen but of interest to give an indication as to just how much information can be obtained from the forecast.

11 *Clouds, Formation and Types; Clouds and Weather*. Charts of cloud types produced for British Petroleum by R. K. Pilsbury, FRPS and available from the Royal Meteorological Society.

12 *Wind and Sailing Boats*, Alan Watts, (David and Charles, Newton Abbot). A good textbook on small scale wind effects such as gusts, bends, squeezes, sea breezes.

13 *Wind Pilot*, Alan Watts, (Nautical Publishing Co. Ltd, Lymington), a climatology of local topographical and diurnal wind phenomena around the coasts of Europe, the Mediterranean and the Baltic.

14 *Admiralty List of Radio Signals, Vol 3*. (Hydrographer of the Navy). Contains full details of weather bulletins broadcast by radio and of radio fascimile broadcasts.

15 *Met. O 509, Ships Code and Decode Book*, available from HMSO. Contains full details of Meteorological Office Atlantic weather bulletins.

Index

CRUISING

BOB BOND

Are you looking for a boat for family holidays? Do you know the legal rules for steering and sailing? Or are you planning a long sea cruise? Whether a beginner or an experienced sailor you will find this book packed with practical advice on every aspect of cruising.

This is not a book of theory but one of practical advice based on the author's own experiences of cruising with his family and friends over many years. The book is divided into two parts: the first discusses preparation – choosing a boat, basic skills, clothing, catering and planning a voyage. The second part looks at the different types of voyage and the particular skills and problems involved – using a marina, day, estuary and coastal cruising, night sailing and using a foreign harbour.

An invaluable guide for the family sailor, this book will add confidence to the basic navigational skills the sea cruiser needs when planning his own cruise.

TEACH YOURSELF BOOKS

NAVIGATION

A. C. GARDNER

Navigation is a rewarding skill essential to all yachtsmen, and anyone who possesses a minimum of mathematical knowledge can master the basic theory and principles from this book.

The author, a highly qualified sea and air navigator, lucidly explains the techniques, instruments and calculations used in all kinds of navigation to give the reader a thorough grounding in the subject. A brief history of navigation is also included and there are many practical examples and exercises.

TEACH YOURSELF BOOKS